George B. Loring

Investigation of the Scientific and Economic Relations of the

Sorghum Sugar Industry

Being a report made in response to a request from the Hon. George B. Loring

George B. Loring

Investigation of the Scientific and Economic Relations of the Sorghum Sugar Industry
Being a report made in response to a request from the Hon. George B. Loring

ISBN/EAN: 9783337418496

Printed in Europe, USA, Canada, Australia, Japan

Cover: Foto ©berggeist007 / pixelio.de

More available books at **www.hansebooks.com**

NATIONAL ACADEMY OF SCIENCES.

INVESTIGATION

OF THE

SCIENTIFIC AND ECONOMIC RELATIONS OF THE SORGHUM SUGAR INDUSTRY,

BEING A REPORT

MADE IN RESPONSE TO A REQUEST FROM

THE HON. GEORGE B. LORING,

U. S. Commissioner of Agriculture,

BY A COMMITTEE OF THE NATIONAL ACADEMY OF SCIENCES.

NOVEMBER, 1882.

WASHINGTON:
GOVERNMENT PRINTING OFFICE.
1883.

LETTERS OF TRANSMITTAL.

DEPARTMENT OF AGRICULTURE,
Washington, D. C., January 10, 1883.

SIR: In compliance with a resolution of the Senate, July 1, 1882, I transmit herewith for the use of the Senate a copy of the report of the committee of the National Academy of Sciences upon the subject of Sorghum Sugar. I deem it proper to state in this connection that I have been unable to comply with the direction of the Senate at an earlier date on account of delay in transmitting the report of the Academy to this Department, which was not received until November 15, 1882; and a communication from the Acting President of the Academy, dated December 23, 1882, which forms a part of the report. The preparations of the voluminous manuscript copy of the report has moreover required a considerable length of time, and the illustrations required by the Academy for the report are just now received, having been supplied to this Department on the 8th instant.

Very respectfully, your obedient servant,
GEO. B. LORING,
Commissioner of Agriculture.

Hon. DAVID DAVIS,
President of the Senate of the United States.

YALE COLLEGE, NEW HAVEN, CONN.,
November 13, 1882.

SIR: I have the honor to transmit to you herewith a report on the sorghum-sugar industry, made by a Committee of the National Academy of Sciences, in accordance with the request contained in your communication of January 30, 1882.

Very respectfully,
O. C. MARSH,
Acting President of the National Academy of Sciences.

Hon. GEORGE B. LORING,
United States Commissioner of Agriculture. ·   3

NEW HAVEN, CONN.,
November 1, 1882.

SIR: Herewith I have the honor to hand to you the report of a committee appointed by the late President Rogers, at the request of the Hon. George B. Loring, United States Commissioner of Agriculture, of date January 30, 1882, for the "scientific investigation of the sorghum question."

The scientific relations of this question are so intimately interwoven with the economic that the two cannot well be considered separately. The Committee were, therefore, glad to learn from a subsequent communication, dated March 24, 1882, from the honorable Commissioner, that "he regarded the investigation of the economic value of sorghum to the sugar manufacturer the point especially interesting to his Department."

Both sides of the sorghum question are, therefore, considered in the report.

Respectfully, yours,

B. SILLIMAN,
Chairman, &c.

Prof. O. C. MARSH,
Acting President of the National Academy of Sciences.

CONTENTS.

PART I.

REPORT OF THE COMMITTEE.

PART II.

CONCLUSION.

PART III.

APPENDED PAPERS.

CONTENTS.

COMMITTEE.

NOTE.—Dr. C. A. GOESSMANN, Professor of Chemistry at the Massachusetts Agricultural College, at Amherst, was also a member of this Committee, and acted in the work until September 15, 1882, when he resigned. The Committee desire to acknowledge the valuable co-operation of their colleague in the inception and progress of this investigation, and to express their regret that his name should not appear on this report, as it so often appears in it, as one of the earliest investigators of the sugar-producing capacity of sorghum.

PART I.

REPORT OF THE COMMITTEE.

NOTE.

JUNE, 1883.

The first draft of this report was submitted to the National Academy of Sciences at its session in Washington in April, 1882. The official copy of the Document was transmitted to the Commissioner of Agriculture in November following.

The Committee have embraced this opportunity to add to their Report the results of the crop grown in 1882, as also some matters of historical interest relating to sorghum.

THE COMMITTEE.

THE SORGHUM SUGAR INDUSTRY.

REPORT OF THE COMMITTEE OF THE NATIONAL ACADEMY OF SCIENCES.

"The sorghum question" referred by the Commissioner of Agriculture to the National Academy of Sciences for investigation and report, means undoubtedly, in the sense in which it now chiefly interests cultivators, the sugar-producing value of the sorghum.

The questions relating to the value of sorghum for food for men or animals, of its use as forage, or for the manufacture of spirits, glucose, beer, and vinegar, &c., are all subordinate to the sugar-producing value of the plant. For more than a quarter of a century sirup has been made from sorghum, over a wide range of country in the United States, both north and south, and for a time it was confidently believed that sorghum culture would assume great importance as a source of cane-sugar. Many efforts were made to establish this industry in various places as a part of the domestic work of the farm. These attempts were rarely other than disappointments. Occasionally, here and there, good crude sugar was made, and it was frequently observed that the sirup, when permitted to stand for a length of time, deposited crystals of cane sugar. But, on the whole, the attempts to manufacture sugar from sorghum, on a scale of commercial importance, were a failure up to the time when the Department of Agriculture took in hand, in its chemical division, the solution of the sugar problem.

That the sorghum was, under certain conditions imperfectly understood, capable of producing cane-sugar, admitted of no doubt, but what the conditions of success were was not known. How confused, contradictory, and ill-digested the state of our knowledge was on this subject, prior to 1878, will be seen from what follows.

Long as the so-called "Chinese sorghum" or sugar cane had been cultivated in China, there appears to be no evidence that the Chinese used it for sugar-making nor even for sirup.*

No data existed in their literature or experience on which to draw conclusions from these ancient cultivators of sorghum showing that they were even acquainted with its sugar-producing nature. This experience has been almost exclusively American and is comparatively recent.†

---

* See in Appendix an interesting statement from the eminent Chinese scholar Rev. S. Wells Williams, professor of Chinese at Yale College, on the so-called Chinese sugar-cane, p. 57.

† The sugar-producing power of the sorghum appears to have been first noticed in France, and our first seed came to the United States from that country. Mr. Leonard Wray, author of the "Practical Sugar Planter," a standard work for a quarter of a century, introduced the sorghum into America from Natal, where he was then resident some thirty years ago. He is also cited by M. L. Vilmorin, as introducing sorghum

A proper introduction to this report, as it appears to the Committee, is a brief sketch of the history of the earlier investigations upon sorghum (especially American examples) as a sugar-producing plant, chronologically arranged.

The following citations are by no means an exhaustive summary, but are, probably, sufficient to show the wide differences of opinion on nearly every important point of the subject entertained by the several authorities quoted.

A few only of the more important points of this inquiry have been selected as illustrations, and the conclusions reached are grouped under each head and chronologically arranged.

CONFLICTING OPINIONS ON ESSENTIAL POINTS.

OF THE KIND OF SUGAR PRESENT IN THE JUICE OF SORGHUM.

a. In a paper by D. Jay Brown (Annual Report Department of Agriculture, 1856, p. 310), he says:

Mr. Hervey, of France, contends that there is no uncrystallizable sugar pre-existing in the cane (sorghum), and that the formation of glucose (grape-sugar) or molasses is only owing to the action of the salts contained in the liquid during the manufacturing process.

b. Dr. C. T. Jackson (Annual Report Department of Agriculture, 1857, p. 187) says:

There is no doubt that this plant (sorghum), when unripe, contains only grape-sugar.

c. Dr. J. Lawrence Smith, in a paper detailing the results of his investigations of sorghum (Annual Report Department of Agriculture, 1857, p. 192), says:

This result settles the question *that the great bulk of the sugar contained in the sorgho is crystallizable, or cane-sugar proper.*

And again, giving his final conclusions, he says:

1. The sorgho contains about 10 per cent. of crystallizable sugar.
2. The sugar can be obtained by processes analogous to those employed for extracting sugar from other plants.

In an article entitled "Contributions to the knowledge of the nature of the Chinese sugar-cane" (transactions New York State Agriculture Society, 1861, p. 785) by Dr. C. A. Goessmann, he says p. 789:

The facts so far obtained prove that, besides cane-sugar, no other kind of sugar exists in the juice of the ripe and sound sorghum-cane.

Again, in describing the general properties of the sorghum-cane juice, he says, p. 798:

I have already mentioned that the results, which I obtained, entitled me to believe that cane-sugar is the only kind of sugar that exists in that juice.

And on page 808, he says of results in extracting sugar from sorghum:

These results are very encouraging, as they show that more than half the sugar, or 5 per cent. out of 9 to 9¼ per cent. in the juice, can be separated. When Achard es-

<hr/>

from Caffraria into France in 1854. (See translation of Vilmorin's paper in the appendix (p. 59). An interesting letter from this veteran promoter of the sorghum sugar culture, of date of September 7, 1882, from Parak, Indian Archipelago will also be found in the Appendix.)

Louis Vilmorin, just mentioned, is the author of one of the early memoirs on this subject (1855), of which a translation will be found in the Appendix, *loc. cit.*

Consult in this connection the bibliography of sorghum in the Appendix, and also an extract from the work on sugar by Mr. Basset quoted on p. 37.

tablished the first beet-sugar manufactory in Silesia, he was able to separate only from 3 to 4 per cent. of sugar, although 10¼ per cent. was present; and the French manufactories were quite contented when they succeeded in extracting from 4 to 5 per cent. of sugar. The history of the development of the manufacture of beet-sugar may be studied with great advantage by those interested in the sorghum.

d. Dr. Thomas Antisell, chemist Department of Agriculture (Annual Report Department of Agriculture, 1867, p. 33), says:

The attempt to separate and crystallize the cane-sugar of sorghum, on a large scale, has been wholly unsuccessful, and, as a sacchariferous plant, it is only valuable for molasses.

e. President Stockbridge, of the Massachusetts Agricultural College, in his Annual Report, December, 1881, p. 19, says:

The experiments with sorghum, as a sugar-producing plant, forever settled the fact that no known variety of it can be profitably employed for the purpose, unless chemical science can discover a law by which glucose can be changed for cane-sugar.

THE BEST VARIETIES OF SORGHUM FOR THE PRODUCTION OF SUGAR.

In the Sorgho Journal for February, 1869, p. 9, the editor, William Clough, says:

The Oomseeana is altogether the best, Neeazana is next, for making sugar. It is not worth while to try to make sugar of any other variety which we now possess.

Again, p. 26, he says:

It [the Oomseeana] is the only cane upon which the operation for sugar can be conducted with any certainty.

Again he says:

Spend no time in attempting to make sugar from any but the Oomseeana or Neeazana varieties.

Again, same page, he says:

Its sirup does not tend to granulate.

TIME FOR HARVESTING AND WORKING THE SORGHUM, AND WHEN THE MAXIMUM OF SUGAR IS PRESENT IN THE JUICE.

a. In the Annual Report Department of Agriculture, 1854, p. 222, M. Vilmorin, of Paris, is quoted as concluding that—

The proportion of sugar in the stalks continued to increase until the seeds were in the milky state. * * * The ripeness of the seeds does not appear much to lessen the production of sugar, at least in the climate near Paris; but in other countries, where it matures when the weather is still warm, the effect may be different.

b. J. H. Hammond, Silverton, S. C., Annual Report Department of Agriculture, 1855, p. 282, found by his experiment (he records one only) with sorghum, taken before the seed was in the milk, when it was in the milk, and when it was mature, that

The youngest canes had rather the most, and the oldest rather the least saccharine matter. * * * Beginning to cut the cane as soon as the head is fully developed, it may be secured for a month before it will all ripen; how long after that, I do not know.

c. Dr. C. T. Jackson (Annual Report Department of Agriculture, 1856, p. 307) found that—

The juice from stalks, with quite ripe seeds, was, by far, the sweetest, while the green one, which was just in flower, contained but very little saccharine matter.

Upon page 312 Louis Vilmorin is quoted as saying:

The crystallization of the sugar of the sorgho, it seems, should be easily obtained in all cases where the cane can be sufficiently ripened; and, as the proportion of the sugar is an unfailing index of ripeness, it follows that we could always be sure of obtaining a good crystallization of juices, the density of which exceeds 1.075, while weaker ones could not yield satisfactory results after concentration.

Again he says, same page:

This difficulty [of purging, through presence of the gum] only presents itself in the employment of unripe canes; for, as soon as the juices attain the density of 1.080 and more, they contain little else than crystallizable sugar, and their treatment presents no difficulty.

d. Dr. C. T. Jackson, in his report (Annual Report Department of Agriculture, 1857, p. 187), says:

A ripe plant yielded a juice of 1.062 sp. gr., which yielded 16.6 per cent. of thick sirup, which crystallized almost wholly into cane-sugar, the whole mass becoming solid with crystals.

And he concludes:

From these researches I am fully satisfied that both the Chinese and the African varieties of sorghum will produce sugar of the cane type perfectly and abundantly, whenever the canes will ripen their seeds.

Again he says:

The unripe canes can be employed for making molasses and alcohol, but, as before stated, will not yield true cane-sugar.

e. The committee of the United States Agricultural Society, appointed to investigate the subject of sorghum, in their report (Annual Report Department of Agriculture, 1857) say:

Where the plant was well matured, the juice yielded from 13 to 16 per cent. of dry saccharine matter; from 9 to 11 per cent. of which was well-defined crystallized cane-sugar. * * * A palatable bread was made from the flour ground from the seeds. * * * Paper of various qualities has been manufactured from the fibrous parts of the stalks.

f. J. N. Smith, of Quincy, Ill. (Annual Report Department of Agriculture, 1862, p. 134), says:

The sirup [from sorghum] will not make sugar if the cane is cut before the seed is in the dough. * * * The crop should be allowed to stand in the field as long as possible, without being in danger of frost.

g. L. Bollman, Bloomington, Iowa, upon page 147, *loc. cit.*, says:

To me it is obvious that the chief requisite for sugar-making from the sorghum canes is their *perfect maturity*, and such maturity is dependent on correct cultivation and late cutting.

h. J. Stanton Gould, in a report on "Sorghum Culture" made to the New York State Agricultural Society, 1863 (Transactions New York State Agricultural Society, p. 752), says:

The seed of the cane [sorghum] continues in the dough for about a week. It is the general impression the cane should be cut during this period, as it is then supposed to have the greatest amount of saccharine matter; at least this is thought to be true of all the varieties except the white imphee, which is usually cut *just as it is going out of the milk or just entering the dough.*

i. William Clough, editor Sorgho Journal, Cincinnati, Ohio (Annual Report Department of Agriculture, 1864, p. 59), says:

The precise period most appropriate for harvesting the cane is when the saccharine properties are fully developed, and before any supplementary action sets in. This will be found to be at the time when the seed at the middle of the panicle is just beginning to harden, or to pass from the fluid or milky state.

Again he says (Annual Report Department of Agriculture, 1865, p. 312):

Until recently the opinion has prevailed that cane for making sugar should be thoroughly ripe; that it could not remain standing in the field too long, provided it escaped the frost; but lately this notion has been somewhat modified. * * * Something like a case for early or premature harvesting has been made out. The matter cannot, however, be considered as definitely settled until the results of the season of 1866 shall have been determined. After the next year it will be fully understood. The precise stage of maturity most favorable for the production of crystallizable sugar, according to the new theory, is just after the seeds are formed and before they begin to harden.

j. Prof. Henry Erni, Chemist, Department of Agriculture, 1865, p. 43, says:

Contrary to my expectations, I found that the expressed sorgho juice of ripe cane, whether neutralized by lime or not, refused to crystallize, for what solidified or granulated after long standing of the sirup, was grape sugar.

And, in a foot-note, he says:

The juice from unripe cane readily crystallized.

k. In a pamphlet entitled, "The Sorgho Manufacturer's Manual," by Jacobs Brothers, Columbus, Ohio, 1866, p. 4, it is stated that—

The cane is in the best state for harvesting when part of the seed is beginning to turn black, or, in other words, *when the seed is in the doughy state.*

l. A correspondent of the Department of Agriculture (Annual Report, 1867, p. 359) says:

I take the sorghum (Otaheitana) when just fairly in bloom. In no case do I allow the seed to mature when I wish to make sugar; but for No. 1 sirup I let the cane mature.

m. The Sorgho Journal, William Clough, editor, February, 1869, p. 26, speaking of Neeazana, says:

Do not mind the panicle; if the juice has a clear, sweet taste, even if the panicle is only in bloom, cut and work the cane.

Again, p. 92, under an article entitled "Immature cane best for sugar," it says:

The theory that cane should be harvested before fully ripe, when designed for sugar, has been further confirmed by the experience of this year. The other idea, that the cane should be fully ripe, was never confirmed by facts.

Page 58 it says:

The weight of evidence, just now, is in favor of cutting as the seed is passing from the milk to the dough state.

Again, p. 73:

Cut the cane as soon as the seeds are formed. * * * Cut the cane as soon as they acquire a clear, sweet taste. This may occur in some seasons when the cane is in the flower, and in other seasons not till the seed is fully formed.

n. E. W. Skinner, of Sioux City, Iowa, says (Annual Report Department of Agriculture, 1873, p. 393):

The best sirup is made from cane not fully ripened.

o. In his report on "Early Amber Cane," by Dr. C. A. Goessmann, of Amherst, Mass., 1879, he says, p. 9:

The safest way to secure the full benefit of the Early Amber cane crop, for sirup and sugar manufacture, is to begin cutting the canes when the seed is full grown, yet still soft.

p. In the Sorgho Hand-Book, published by the Blymyer Manufacturing Company, Cincinnati, Ohio, 1880, it is directed, upon p. 8:

The cane should be cut *when the seed is in the dough.*

q. In a "Report on the manufacture of sugar, sirup, and glucose from sorghum," by Professors Weber and Scovill, of the Illinois Industrial University, 1881, p. 22, they say:

The proper time to begin cutting the cane is *when the seed is in the hardening dough.*

r. Vilmorin, of Paris, in the *Journal d'agriculture pratique,* February 17, 1881, p. 230, says:

The period during the development of the plant (sorghum) when the juice is purest and richest in sugar is that which precedes the maturity of the seed. It is at that point when the interior of the seed has the consistence of soft dough, easily crushed under the finger-nail, that the plant should be cut and pressed.

PROMPT WORKING OF THE SORGHUM AFTER CUTTING.

a. Dr. J. Lawrence Smith, in his report (Annual Report Department of Agriculture, 1857, p. 192) says:

The uncrystallizable sugar forms rapidly after the cane is fully ripe and recently cut.

And again, as the result of his examinations, he says:

Hence it is evident that no time is to be lost, after cutting, in expressing the juice.

b. D. M. Cook, Mansfield, Ohio (Annual Report Department of Agriculture, 1861, p. 311), says:

Let the cane fully ripen if possible. If the cane is fully ripe it may be worked into sirup and sugar with advantage as fast as it is cut up; but if the juice is not perfectly matured, it should be allowed to "season" a few days. [By having the cane cut up, bound in bundles, and shocked under a barn or shed for a few days.]

c. In an article on "Sorghum culture and sugar making," by I. A. Hedges (Annual Report Department of Agriculture, 1861, p. 297), he says:

After the canes have been topped, stripped, cut up, and tied in bundles, they may be set up in the open air, or, preferably, under shelter, and kept for some weeks. Such keeping improves the juice not only in flavor, but also in saccharine richness from 1 to 3 degrees B. This improvement takes place upon the same principle and from similar causes which determine the sweetening of acid fruit after pulling, viz, the change of gum and starch into sugar.

d. J. H. Smith, Quincy, Ill. (Annual Report Department of Agriculture, 1862, p. 134), says:

The cane should be cut and brought to the mill and crushed on the same day; and the topping of the cane and the stripping of the leaves from the stalks should proceed no faster than it is cut and brought to the mill, if the very best results are desired and all danger of souring is to be avoided. * * * It is much better, therefore, not to give the cane any rest, after being stripped and topped, till the juice is expressed and run into sirup. * * * When the cane is ripe, it should be immediately cut, for if suffered to remain, after it is ripe, in connection with the roots, a deteriorating effect upon the quality and flavor of the sirup will be the result, and at the same time the quantity will be greatly diminished.

e. William Clough, editor of the Sorgho Journal, says (Annual Report Department of Agriculture, 1865, p. 312):

It would be best to allow but little time between harvesting and working the cane, and on no account should it be stored and allowed to remain long in large shocks. It is almost demonstrable that no cane sugar is developed under any circumstances after the cane is harvested. The changes that occur after the cane is cut, if any, must be in their nature depreciative, consisting in the transformation of crystallizable to uncrystallizable sugar.

f. The Sorgho Manufacturer's Manual, Jacobs Brothers, Columbus, Ohio, 1866, p. 4, directs that:

The cane should be cut and shocked in the field, with tops on, and in this condition it may remain several months before being worked up, for the cane matures and forms more saccharine matter.

g. A correspondent (Annual Report Department of Agriculture, 1867, p. 359) gives his method of working:

I strip, cut, and work up the cane the same day, if possible.

h. E. W. Skinner, Sioux City, Iowa (Annual Report Department of Agriculture, 1873, p. 393), says:

As soon as matured, cut, pile, and cover with leaves; never allow it to stand, after maturity, in connection with the roots.

i. The Sorgho Hand-Book, Blymyer Manufacturing Company, Cincinnati, Ohio, 1880, p. 8, directs that:

The cane should be cut several days before grinding, as it will be more free from impurities if cured for a few days before going to the mill.

j. Professors Scovill and Weber, in their report, 1881 (Illinois Industrial University), say:

The cane (sorghum) should be worked up as soon as possible after cutting.

THE NECESSITY OF FURTHER INVESTIGATION OF SORGHUM.

a. D. J. Brown (Annual Report Department of Agriculture, 1856, p. 313) says:

Let the same skill, directed by science, be applied to the making of sugar from the sorgho sucré, and we may reasonably expect the happiest results.

b. Prof. J. Lawrence Smith (Annual Report Department of Agriculture, 1857, p. 192) further says:

On investigating the sugar-bearing capacity of the Chinese sugar-cane, the first step required was to ascertain the true chemical constitution of the juice extracted from the plant. From various conflicting statements on the subject nothing satisfactory could be gleaned, some of the best authorities insisting that there was not any crystallizable sugar in the juice, or but a very small portion, while others, equally as strong, held the contrary opinion.

c. Dr. J. Lawrence Smith (Annual Report Department of Agriculture, 1857, p. 192) further says:

It must not be forgotten that sugar-making is an art, and cannot be practiced by every one with a mill and a set of kettles. * * * What was necessary for the beet root is doubtless required for the sorgho, namely, a thorough study of its nature, with a process of extracting the sugar specially adapted to it.

d. J. Stanton Gould, " Report on sorghum culture," (Transactions New York State Agricultural Society, 1863, p. 740) says, in view of the discordant testimony concerning the sorghum question :

These conflicting opinions might easily be reconciled by a few well-directed experiments.

Again he says, same page:

After the most careful inquiry, orally and by letter, I am unable to find that any such experiments have ever been made.

Again he says (p. 747:)

These experiments are not conclusive, and the whole question needs a careful and accurate investigation.

e. Dr. J. M. Shaffer, Secretary Iowa State Agricultural Society, says (Annual Report Department of Agriculture, 1868, p. 515) :

The production of sugar (from sorghum) is rather the result of accident than of any well-digested system for its extraction.

From the foregoing discordant statements upon some of the more important points selected for comparison, viz, (1) the kind of sugar found in the sorghum; (2) the best variety of sorghum for the production of sugar; (3) the time for harvesting and when the maximum of sugar is present in the juice ; (4) the prompt working of the canes after cutting, &c.; it is evident that nothing was definitely determined even on points where work in the laboratory and the exercise of analytical skill was apparently sufficient to settle most doubts, aside from economic questions, relating to methods of manufacture.

Such we find was the condition of the " sorghum-sugar question " up to a period immediately preceding the researches undertaken by the United States Department of Agriculture in 1878 by their present chemist, Dr. Peter Collier.

All the analytical and scientific work of this chemist has been before us, either in the published reports of the Department of Agriculture; in the manuscript report of the work of 1881 and '82, with its appendices, submitted to the commitee by the honorable Commissioner of Agriculture, supplemented by the personal inspection of the chemical methods in use in the laboratory of the Department in Washington by members of the Committee; by the examination of Dr. Collier before the Committee in more than one session at New Haven, where he was invited for this purpose; and, lastly, by correspondence during the whole period covered by the work of this Committee.

To secure the results of other chemists and workers in this field of research, correspondence has been opened by this Committee with those who responded to circulars sent out asking for co-operation and information on the sorghum question.

The results of these inquiries will be found appended, with acknowledgments to those who have so efficiently aided the work of the Committee on both the scientific and economic side of this investigation.

A full digest and analytical summary of the present state of our knowledge of this subject is presented in Part II of this Report.*

It will be observed that the existence of sugar in the stalks of maize is frequently mentioned in the several reports of the Department of Agriculture, and comparative statements are made between the sorghum sugar results and those obtained in a parallel series of experiments conducted at this Department upon maize.

This subject was not specifically referred to this Committee, nor have they devoted much time to its consideration. It was, however, found convenient to give the results in brief, for what they may be worth, and without expressing an opinion on their practical value.

The analytical methods of investigation employed were the same with those used in the investigation of sorghum. There appeared to be no sufficient reason for omitting these comparisons, which are intimately woven into the text of the several documents before us. Whether these results may or may not be reproduced in field culture on a large scale and with commercial success are points requiring further experimental tests, and on these points the Committee are not now prepared to express any opinion.

THE AGRICULTURAL CHARACTERS OF SORGHUM.

The cultivated varieties of sorghum, considered botanically, are cereals. They belong more especially to that very small group of cereal species which have been cultivated from the dawn of history and have developed along with our civilization. During ages of culture they have so changed under the hand of man that we are ignorant as to their native countries, and know not what their original wild progenitors were. Their descendants now exist in a vast number of varieties, which differ so greatly among themselves that neither scientific botanists nor practical cultivators are agreed as to what are true species and what mere varieties which have arisen in cultivation.

The cultivated varieties of sorghum have been placed in the genera *Holcus*, *Andropogon*, and *Sorghum* by different botanists, the latter being the name now accepted.

---

* For a fuller notice of the literature of sorghum, reference is made to a "Bibliography of Sorghum," which will also be found in the appendix to this report.

The Committee are aware that many titles of memoirs, scattered through a wide range of periodical literature in various languages, might be added, but it has not been in their power to make the search required to complete the list. The librarian of the Department of Agriculture has rendered efficient aid in compiling this bibliography.

A generation ago botanists grouped the numerous cultivated varieties into a considerable number of distinct species, without agreement as to how many; five or six were generally believed to exist. Certain varieties of durra, with the grain in a somewhat loose panicle, and which were more especially cultivated in Asia and in Southern Europe, were classed as one species called *Sorghum* (*Holcus* or *Andropogon*) *vulgare;* the varieties with the grain in a densely contracted panicle, grown more largely in Africa, and known as Guinea-corn, Egyptian durra, Moorish millet, &c., were grouped into another species, called, *S. cernuum;* the variety best known as chocolate-corn was the *S. bicolor;* broom-corn and all the sugar-producing kinds were classed together as *S. saccharatum;* and other specific names were applied to smaller groups of these varieties.

But the investigations of modern science have gradually led to the belief that all the numerous varieties once classed in the several species above enumerated had a common origin and constitute but a single species, to which the old name *Sorghum vulgare* is now applied. This is now the belief of the most eminent botanists of the world. Some even go further and believe that *all* the cultivated varieties of the genus, including the spiked millets (*Sorghum* (*Holcus*) *spicatum*), are the descendants of a single original parental species.

These conclusions have a most important bearing upon the subject of this special investigation.

It is a law of nature that the longer a species is cultivated and the wider its cultivation extends the more easily it changes into new varieties and the wider the differences between the varieties become. Some species, however, have a much greater capacity for variation than others, and *Sorghum vulgare* stands pre-eminent among the useful plants for this character.

The usefulness of any agricultural species is intimately correlated with its capacity for variation in cultivation, for this means capacity for the improvement of varieties by the only means known to cultivators by which such improvements may be effected. It also means capacity for adaptation to varied conditions of soil, climate, and natural surroundings, and, furthermore, adaptation to various methods of culture and to various uses. It is a sort of plasticity which allows the species to be molded in the hands of the intelligent cultivator.

This species (*Sorghum vulgare*) has varied more widely under cultivation than any other cereal, unless it be Indian corn. The varieties differ in all their characters, in height, fruitfulness, habit of growth, grain, stalk, leaf, panicle, chemical composition, preference of soil, climate, and exposure, and so on to all the differences in which species themselves differ. Its cultivation has extended to most of the warm and many of the temperate climates of the globe, and it has adapted itself to the varied uses and more varied agricultural methods of nearly all the civilized races of mankind.

The agricultural success of any plant in a country depends in part upon its fitness to the soil and climate, and in part to a variety of other conditions, one of which is that it must fill some place in the agriculture of that country better than the other species competing with it. Sentiment and local customs are also factors, but which have less force in this country than in others.

Durra, Guinea-corn, broom-corn, and probably also chocolate-corn, were introduced into this country in colonial times. During the days of more imperfect tools and machinery, and of difficult transportation, all our agricultural crops were of necessity grown upon a much smaller scale than now, and on most farms a greater variety of crops were grown than now. Most if not all the agricultural plants of the Old World were tried here, and many had a wide and sparse cultivation until well into the present century, and then disappeared under the new conditions of our agriculture. The cultivation of others became specialized. Varieties of this species may be found in both these categories. Durra and Guinea-corn were widely introduced, and they lingered in cultivation until crowded out by Indian corn. They were dropped just as many other minor crops were; they did not fill a place in our modern agriculture so well as some other species did, and now are only found in regions where Indian corn does not grow so well, particularly in the States which border on Mexico. Chocolate-corn (the old *S. bicolor*) was cultivated here and there as a poor substitute for coffee, but under the changed conditions of things it has entirely disappeared from our fields and gardens, crowded out by imported and better coffee. Broom-corn, also introduced in colonial times, was widely cultivated; forty years ago very many persons grew enough for their own use or for local sale. It supplied a certain want better than anything else, consequently it could not be crowded out, but under the conditions of modern agriculture its cultivation has become specialized and concentrated in fewer localities, in some of which it has assumed an importance found nowhere else in the world. It has been greatly improved, and the cultivation of American varieties has now extended to the Old World.

About thirty years ago the sugar-yielding sorghum was introduced. Filling a certain place on our farms better than any other plant previously tried, it spread in cultivation with a rapidity no other agricultural plant ever did before in this or any other country, and is the only one adapted to a wide region introduced into the United States since colonial times which has become of sufficient importance to be enumerated in the census. It has become the "sorghum" of common language, and its cultivation has extended the whole length and breadth of the country.

Its adaptation to our soil and climate is abundantly demonstrated, and its capacity for improvement also thoroughly proven. . The Department of Agriculture has already examined more than forty varieties, some of which have originated in this country. We have now varieties with very unlike characters; some mature in 80 days, others require twice as long a time, and one variety has become in a sense perennial, a fact not true of any other cereal species grown in the country. They vary in habit of growth and in sugar-content; the two extremes have been developed here—the one as rich as Louisiana sugar-cane, the other the broom-corn, so poor in sugar.

Belonging to such a plastic species, with such adaptation to a wide range of soil and climate, with such capacity for modification and improvement, already in such wide cultivation, and promising to meet such a definite want in our agricultural production, it is certain that, in obedience to natural laws, some of the existing varieties may be greatly improved, and that new ones may be made, some of which will better serve the ends we are now seeking than any varieties we now have. No efforts have yet been made to increase the sugar-content by systematic, intelligent, and long-continued selection. In the light of the successful results of experiment in this direction with sugar-beets, and with the abundant experience we have with other species as to other results attained by such processes, we have much to hope as to improvement in this character with a species which has been so variously molded to the uses of man.

Agriculture, however intelligently pursued, is more of an art than a science. Hence the ultimate profitableness of any agricultural crop introduced into a region new to it can only be determined by actual trial through a series of years. The nature of the economical problem is such that science cannot predict the result. It can, however, render great aid in making success more probable, and in hastening it where it otherwise might be much delayed. It can suggest means and methods, can indicate promising directions for experiment, can aid in foreseeing and overcoming many difficulties, suggest remedies for mishaps, and in a multitude of ways aid in solving the practical problem. This is especially true when the crop is to be manufactured into a commercial product, and emphatically so in the production of sugar, the whole economical aspects of which have been changed by the aid of modern science.

No agricultural species can be cultivated profitably everywhere within its range of actual growth, and it is yet to be demonstrated where the best regions are for the most profitable growth of sorghum. This is only partly an agricultural problem; it is as intimately related to the question of winning the sugar in the best form and at the least expense. For the solution of the latter, scientific work is needed. It can ultimately be done in the sugar-house; it may be more quickly done, and with vastly greater economy, if this be aided by the scientific laboratory. The profitable production of sugar from cane, as now pursued in Louisiana, and from beets, as pursued in Europe, was achieved only by such aid. The methods of extracting sugar from these two great sources are very unlike, and each was developed along with scientific investigation instituted for each special plant. Sorghum still needs this. The work so nobly begun and successfully pursued by the Agricultural Department is still incomplete and unfinished. To use an agricultural simile, the crop has been sown, but the harvest has not been reaped.

Agriculturally, the sorghum question is solved, so far as it can be until science now does her share. That the crop may be widely and economically grown, containing a satisfactory amount of cane-sugar, is sufficiently proved. All the problem remaining unsolved relates to the extraction of sugar. In view of the magnitude of the interests involved, the results already obtained, and the wide attention the matter is now receiving, we feel that there are most encouraging indications of practical success.

VALUE OF THE RESEARCH IN A MATERIAL SENSE TO THE NATION.

Aside from the value of this research from a scientific standpoint, illustrating as it does the importance of obtaining from an extended investigation the facts and their mutual relations in an agronomic problem, the results obtained appear to this Committee to possess a high value, in a material sense, to the nation.

Whether the cultivation of a crop like sorghum, deriving its support largely from the atmosphere and water, since it appears to thrive best upon light soils, may or may not reward the cultivator better than the growth of cereals, it certainly adds a new factor to agriculture, of value not only as a sugar-producing plant, but also as a food plant of no mean quality. It thrives over a very wide area, and, as we have shown, develops in the warm and temperate latitudes more than a single crop per annum, and becomes, certainly in one of its varieties, perennial.

But the work is also of national importance in its relation to existing industries, and especially to that of the cultivation of the sugar-cane and sugar-production therefrom.

In this country the sugar-planter has to contend with obstacles unknown to the resident of tropical countries. A greater degree of skill and knowledge is here required for the attainment of the same result that elsewhere is reached through the normal operation of natural causes, almost without effort on the part of the planter. Such skill and knowledge can only be attained by a carefully conducted experimental inquiry, such as this investigation exhibits.

The methods developed in the course of this investigation are also applicable, with but slight modification, to the cultivation of the sugarcane, and there can be little doubt but that the ultimate effect of such investigations will be to stimulate the Southern sugar-planter to similar experiments for the ascertainment of the most favorable conditions for the prosecution of his own special industry, depending on the culture of tropical cane in subtropical climates, where it never attains its fullest development, and is consequently subject to many adverse conditions unknown in the tropics.

As a work of national importance, calculated directly to benefit widely separated sections of the country, it is one that has been wisely undertaken and encouraged by the Department of Agriculture, and is deserving of every aid that Congress may be willing to grant for its encouragement and prosecution.

The sugar-planter of Louisiana and Texas may possibly discover that he has at command, in one or more of the larger varieties of sorghum, which, like the so-called "Honduras," "Mastodon," &c., attain at maturity, say in four or five months, a growth of 18 to 20 feet in height, and a weight of 2 to 5 pounds per stalk, a sugar-producing plant thoroughly adapted to his climate and soil, equal, and possibly superior, in productive capacity of cane-sugar to the "Ribbon," "Red," or "White" cane now grown there, and escaping the perils from frost which always attend the cultivation of the cane in those regions where the season is never long enough to permit its full maturity.

Of the early maturing varieties, like "Early Orange," it will be possible in southern latitudes to make two crops of sugar and seed in one season, and these, alternating with varieties of longer periods, may extend the sugar crop over nearly half the year.[*]

FURTHER INVESTIGATIONS DESIRABLE.

The important work of the past four years at the Department of Agriculture, while it has made substantial additions to our former

---

[*] We cite in this connection the following letters from Col. H. B. Richards, of La Grange, Texas, the first to Mr. Hedges, and quoted in his communication in Appendix XV:

"But now let me tell you about my *Orange cane.* It is no longer doubtful at all but that the *Orange cane* will become in this climate perennial, and after this year I will only plant every two years. I have tested it now effectually for two years, and am convinced that the stubbles will stand colder weather and more of it than those of the Ribbon cane.

"My cane from last year's stubbles has larger stalks, is taller, and in every way ahead of the earliest seed cane at this time.

*   *   *   *   *   *

"Yours truly,

"HENRY B. RICHARDS.

"LA GRANGE, FAYETTE COUNTY, TEXAS, *April* 8, 1882."

Also his communication to the Chairman of this Committee, given in full in the Appendix (p. —), of date September 25, 1882, in which he adds the experience of the present season, substantially confirming and extending his former statements.

knowledge, leaves yet much to be accomplished in the same general direction.

To be of practical utility to the farmer, the work of the laboratory must be put to the test of experience, that the principles deduced from research in the small way and in a single or a limited number of seasons may be extended to meet various and possibly unfavorable conditions, and over a sufficient area and period of time to permit of a careful and thorough investigation under fairly average conditions.

Among the more important points yet to be investigated may be mentioned the ascertainment by direct working tests on a manufacturing, or at least a large experimental, scale of the relation existing between the actual manufacturing yield and the proportion of *available sugar*, deduced from the analysis in accordance with the results of our previous experience with the juice of the sugar-cane and beet. This point is essential to the realization of the greatest practical benefit from the work already accomplished in the course of this extended investigation.

A large number or new sorts of sorghum from China, India, and Africa, have lately come to hand wholly unknown, and among them many important varieties never before on trial.

The whole subject of the best methods of defecation, the use of lime, of sulphurous acid and the bisulphites, of strontia, the affusion of cold water, and other untried means, is in a state requiring further examination and experiment before the important conditions on which much of the success of the sugar industry depends can be properly settled.

The question of the use of fertilizers, what they should be, and how used, is in an unsatisfactory condition, as can be seen by reference to the results at the Department of Agriculture, those at the New Jersey Agricultural Experiment Station, and also those of Professor Swenson in 1882, reported in the Appendix.

Here also, at the Department of Agriculture, alleged improvements in methods of culture and manufacture, which are liable not infrequently to mislead the people, occasioning disappointment or loss, can be tested with an authority which is found only in the impartial conclusions reached by official examination or experiment in competent hands.

The Department of Agriculture, with its varied resources, scientific skill, mechanical appliances, and extended correspondence, coupled with the enormous circulation of its publications, can do this work as it cannot be done elsewhere.

THE ANALYTICAL METHODS EMPLOYED.

The Committee, after a careful examination of the analytical methods employed by the Chemical Division of the Department of Agriculture, find that they are entirely sufficient for the work to be done. The details of the processes for the volumetric determination of sucrose and grape-sugar are fully exhibited on pages 9–11 of Special Report 33, and in the Annual Report of 1879, pages 66, 67.* These methods have been skillfully adapted to the character of the proximate constituents of the complex juices to be analyzed, and are among the best known to science.

---

* The limits of error, as shown to the Committee from a considerable number of unpublished determinations, sustain the conclusion that the method employed for the estimation of cane and grape sugars was exceptionally accurate, and more subject to a minus error of 0.2 per cent. on a 10 per cent. solution of pure sugar than to a plus error.

These methods have been employed with precautions adapted to the exigencies of the special problems for the solution of which the investigation has been instituted. By a judicious system of checks and control, and by the reduction to the lowest limit of the personal error of the observer, the accuracy and constancy of the results have been assured as far as, in the present state of our knowledge, such end can well be attained.

The care with which the methods for the determination of cane-sugar have been tested, and the probable error determined, enlists our confidence. The reserve with which the Chemist has refrained from accepting the results as conclusive, until by repetition and variation in the methods he had exhausted the means at his command to prove them to be erroneous, is in the true spirit of scientific research.

The analytical work prior to 1882 comprises the enormous number of nearly 4,500 analyses of forty varieties of sorghum and twelve varieties of maize, covering all the later stages of development of the growing plant. The average results of these analyses, conspicuously displayed in the form of graphical charts, afford a connected view of the progressive development of the juice through the various preliminary stages to and beyond the point of complete maturity.

Such an amount of analytical work as is implied in the careful conduct of nearly five thousand quantitative analyses, with a rather limited number of assistants, and in an inconveniently arranged and contracted laboratory, was rendered possible only by the most rigid system and subdivision of labor in the work—a system in which each assistant was, for the time, devoted exclusively to one thing, e. g., determinations of density by the balance, volumetric determinations of glucose and sucrose, polarizations, ash determinations, total solids, ash analyses, analyses of the seed, quantitative determinations of acids and other proximate constituents of the juices at seventeen different stages of growth of the plant and after maturity. By this system each coworker became thoroughly expert as a specialist in his own duty; and it was thus possible by this system to test the accuracy of the work by submitting identical samples in duplicate and separate numbers for analysis by the same and by different coworkers—a crucial test of verification.

The Committee have critically examined the work done in this way, and for the details, showing a surprising agreement, refer to the appendix (p. 142).

COMPARATIVE RESULT OF ANALYSIS AND POLARIZATION.

The optical method of determination of sugar values, now commonly employed by sugar boilers, has found a wide term of comparison with the analytical results in the sorghum and maize sugar researches of the Department of Agriculture.

The comparisons in 1879–1881, between large numbers of determinations by the cuprous precipitation and by polariscope appeared to sustain the opinion that there was a pretty constant difference in favor of the volumetric method, i. e., that the polariscope for some unknown reason failed to detect as much sugar as was demonstrated by the method of precipitation. These differences are set forth below, together with the very satisfactory results of over five hundred similar determinations made in 1882, from which it clearly appears that the discrepancy formerly noticed is apparent and not real. This conclusion removes any doubt which hung over the practical value of the optical method; and this is practically of much moment, for in the rapid operations of the sugar plantation, during the pressure of the crop, the polariscope is nearly the sole dependence of the superintendent in judging many times daily how his juices are running.

In 1879 this comparison was, between sorghum and sugar-cane, as follows:

	Number of anal	Average sucrose by volumetric analyses.	Average sucrose by polariscope.
		Per cent.	Per cent.
Sorghum	22	13. 26	13. 15
Sugar-cane	6	13. 30	13. 09

In 1881 the number of these comparisons was very greatly increased, being between 697 analyses of sorghum and 103 analyses of maize.

Calling the value of the sucrose, as found by analysis, 100, the value indicated by the polariscope was 94.87 for the maize, and 95.96 for the sorghum. The nearly constant difference of about 4 per cent. less sucrose, as determined by these polariscope tests, than was found by cuprous precipitation, was, for the time, attributed to a portion of invert sugar, and to various causes which probably were misconceptions, seeing that this discrepancy disappears almost entirely in the results of the present year, viz: Number of analyses and polarizations 517, of some forty varieties of sorghum.

Total polarization, 5,440.76; average percentage, 10.524.
Total by analysis, 5,433.72; average percentage, 10.510.
10.510: 10.524 = 100: 100.13.

Each result of the 517 is of record, but the general result given suffices. The conclusion seems justified that any differences existing in the polarization and analyses with *normal* fresh juices are only differences incidental to the work, and are not caused by any active rotatory substance present other than *sucrose*. If the juice is *abnormal*, very wide differences may exist. This was conspicuous in the mill work at the Department in 1881, both in juices and sirups.

SUMMARY.

The facts relating to the economical production of crystallizable canesugar on a scale profitable to the farmer and manufacturer, from sorghum, in this country, so far as developed by the existing state of laboratory and field practice, appear to the Committee to be as follows, viz:

1st. That these plants develop at maturity, and when the seed is ripe, a maximum of cane-sugar and a minimum of glucose.*

2d. That the maximum of cane-sugar in sorghum juices is found associated with about one-tenth its weight of grape-sugar (glucose), and not far from one-fifth its weight of "solids not sugar," viz, ash, gum, chlorophyll, albumen, wax, aconitic acid, &c.

3d. That after maturity the relative amounts and proportions of the

---

* Some of the widely discrepant statements by different observers may find, in part, an explanation from the fact developed by the late investigation of Dr. Collier, that in sorghum the sucrose appears to fall off or come to rest during the ripening of the seed, and then again after to increase. This interesting point has been fully developed only by the work of the year 1882, the full details of which will be found in the Annual Report of the Department of Agriculture for 1882.

chief factors vary but little, even for a period of three months* or more, provided the season does not change; *e. g.*, an early maturing variety of sorghum holds its own until frost; a later variety has a shorter working period.

4th. That while varieties of sorghum differ greatly in rapidity of growth and time of reaching maturity, in size, weight, and consequent yield per acre, it appears that all varieties of sorghum resemble each other in developing at maturity, under the same conditions, nearly the same maximum percentages of cane-sugar, glucose, and solids, the cane-sugar maxima varying from 14 to 16 per cent. of the total weight of the expressed juice, the other factors being as stated under 2d.†

5th. The soil best adapted to the growth of a good crop of sorghum for sugar appears to be a sandy loam.‡ This plant thrives on soils and in climates too light and dry for maize, and makes the best "stand" when grown closer than Indian corn admits in a given locality.

6th. While good sirup may be produced from sorghum as a domestic industry and on a limited scale over a very wide range of country, the successful production of 'crystallized sugar on a commercial scale appears to demand the skill and appliances of a sugar-house conducted in a systematic manner and with ample capital.

7th. The best results in sugar are obtained only when the ripe cane is manufactured on the same day (twenty-four hours) in which it is cut from the field.

8th. The seed of ripe sorghum is a valuable feed crop, comparable for fattening animals with maize, and in product is equal to from $2\frac{1}{2}$ to 4 bushels per ton of cane.

9th. About 40 per cent. of the juice of sorghum is lost in the begasse, as it is to nearly the same extent in tropical sugar-cane, more than one-half of which loss may possibly be saved to the crop by processes under investigation.

10th. Of other residual products, the scum and sediment, rich in various elements of fertility, are now thrown away. (For the constituents of these waste products see the analyses at foot of page 29.)

The begasse, when treated by a pulping machine, gives a valuable paper stock. Treated as a fertilizer, the begasse will return to the soil a portion of what the plant has borrowed from it in its growth. In regions where fuel is dear the begasse can be used with advantage as fuel.

---

* A longer working period than three months has developed itself by the experience n Texas upon the "Orange" variety. See letter of Colonel Richards of date September 25, 1882, in the Appendix.

† This generalization appears fully justified by the work done at the Department of Agriculture and for the latitude of Washington; but it is yet an open question how far different sorts of sorghum may vary with climate and soil, two factors of commanding importance as yet imperfectly known.

‡ For considerations of soil and climate, as well as fertilizers, reference may be had to the text, where these subjects are discussed, as well as to various statements in the Appendix.

PRODUCTION OF SUGAR FROM SORGHUM—FAILURE AND SUCCESS.

Repeated failures in the cultivation of sorghum for crystallized sugar as a commercial undertaking had naturally produced distrust of all attempts to renew an industry attended already by many disappointments. It is not, therefore, without reason that some decided successes in making sugar from sorghum on a large manufacturing scale should be demanded before these unfavorable convictions should yield to new evidence.

Considering the former discordant and unsettled state of opinion on this subject, as already set forth in the opening of this report, we can hardly wonder that failure was the rule and success the exception in the former attempts to produce sugar from sorghum. The juice of sorghum even in its best state of development is an extremely delicate and unstable solution of sugar, passing rapidly from sucrose to glucose under the influence of various factors which act to transform it, unless manipulated with skill and in suitable apparatus. These conditions are rarely met at the hands of the small or unskillful cultivator or manufacturer. Hence sirup and not sugar was the result in the great majority of attempts at sugar making; a result by no means without considerable value to the farmer, however unsatisfactory to the sugar boiler. These negative results in the light of our present knowledge and experience prove nothing but the want of attention to conditions essential to success.

FAILURE AT THE DEPARTMENT OF AGRICULTURE IN 1881.

The failure to obtain not only sugar, but even a reasonable quantity of sirup, from the sorghum crop planted in 1881 for the Department of Agriculture on about 135 acres of land near Washington, is an illustration of the importance of adhesion in practice to the principles developed in the laboratory.

It appears from the full statements of the Report of 1882 that, owing to various causes, much of the Washington crop was three times planted,

Analyses of sediment and scum of sorghum in sugar making.

	Sediment.	Scum.
	Per cent.	*Per cent.*
Ether extract, wax, fat, chlorophyl, &c.	16.28	9.53
Alcohol extracts, sugars, resins, &c.	8.06	27.00
Water extract, gum, &c.	33.81	38.83
Insoluble in ether, alcohol, and water	40.86	23.98
	99.01	99.34
Ash, per cent	19.49	13.07
Potash (K_2O)	12.36	19.81
Soda (Na_2O)	3.87	6.03
Lime (CaO)	32.13	26.43
Magnesia (MgO)	2.42	1.92
Sulphuric acid (SO_3)	1.04	2.62
Chlorine (Cl)	2.34	6.02
Phosphoric acid (P_2O_5)	6.18	2.39
Silica	27.81	23.40
Sand, &c.	10.01	10.93
	98.16	99.55
Nitrogen, per cent	2.55	1.46

These analyses are from Dr. Collier.

the last planting being after the middle of June,* thus producing a very imperfect crop, little, if any, of which was in a fit state to be cut and manufactured.

On this point the statements of Peter Lynch, the sugar boiler, are conclusive, there being, as he says, but two days, October 4 and 5, when he received cane in even a reasonably mature state, and from this he readily produced sugar. The report of Assistant Parsons, who had immediate charge of the chemical and other work in the mill, will be read with interest as a conclusive statement of the several causes of failure made by an expert of ample experience.

The insignificant quantity of seed obtained from 93½ acres of sorghum (the other 40 odd acres were too immature to be cut before severe frost), viz, 150 bushels, or 1⅔ bushels per acre, is sufficient evidence of the immaturity of this crop as a whole, and sufficiently explains why even the sirup fell far below the normal quantity.

A reference to the statement of Professor Cook, of New Jersey, State chemist (see Appendix 1, p. 74), shows that the yield of seed from 700 acres of sorghum in that State in 1881 was, even from an imperfect crop, 20 bushels to the acre.

COMPARATIVE FAILURE OF THE FARIBAULT WORKS.

The statement of Captain R. Blakely, of Saint Paul, is of interest in this connection. In his letter to this Committee, of date April 18, 1882, he states the results obtained by him at the Fairbault works, very imperfectly constructed and disadvantageously placed for the delivery of cane, which were commercially a failure, although producing some 15,000 pounds of good sugar, samples of which he has placed in our hands. This witness states his conviction that sorghum sugar—

Is to be one of the great industries of this country. * * * If it can have the fostering care of the Government until it can be established, it will astonish the country.——

SUCCESSES.

One signal success, on a large scale, obtained by intelligent attention to the results of experimental research and skillful culture, opens the way to a repetition of like results. Among the following examples are several of an equivocal nature, presented simply as illustrative of the importance of observing closely the conditions essential to success, as now made clear to cultivators by the researches of Dr. Collier, but which before this time were imperfectly understood or very badly applied even by fairly intelligent operators.

---

* "Notes on sorghum planted on Dr. J. W. Dean's farm 1881," are in the hands of this Committee, being a diary of his work in planting and cultivating 44 acres of sorghum for the Department of Agriculture. By this record it appears that the Honduras planted May 14, 15, and 16, was replanted June 2, and that he, "June 18, began second replanting of Honduras," and "June 20, finished second replanting of Honduras," and June 20, "began second replanting of Early Orange"; June 21, "finished Early Orange 10 a. m., and began second replanting of Liberian"; "June 22, finished replanting Liberian; June 29, used Early Amber in replanting a few rows of Liberian and began second replanting of Neazana."

Also Mr. L. J. Culver, the farmer who sowed about 60 acres of sorghum for the Department of Agriculture in 1881, makes the following statement of his planting:

"On Tuesday, May 10, commenced planting, using Link's Hybrid and Early Amber seed; planted about 30 acres of each variety, but very little of it sprouted, owing to the cold damp weather that immediately followed the sowing. On May 27, commenced replanting the same varieties. This lot of seed was nearly all destroyed by worms. June 7, commenced replanting the third time, and finished the work June 13. The third lot of seed was rolled in coal tar, in order to drive away the worms. It sprouted quickly, but on July 15 cane did not average one foot in height. Commenced harvesting on September 19." (See Agricultural Report for 1882.)

It is from the States of New Jersey and Illinois that we are able to cite examples of success on so large a scale and attended with such a satisfactory result as fairly puts to rest any doubts as to the production of sugar on a great scale, in a northern climate, with a commercial profit. Our first knowledge of the New Jersey enterprise came from the last year's report of Professor Cook, director of the Agricultural Station, but for the current year we are able to record the personal observations of several members of this Committee and others.

We cite as follows from Professor Cook's report, 1881:

1st. *The State of New Jersey* has, at its Experiment Station, made, during the year 1881, a series of well-conducted trials in sorghum culture and sugar production, the full details of which will be found in the report of the director, Professor George H. Cook. (See Appendix, document, p. 71.)

From this document it appears that, during the autumn of 1881, the sorghum cane of 700 acres of land was worked for sugar at the sugar-house of Charles M. Hilgert. This sugar-house has been built at Rio Grande, Cape May County, New Jersey, at a cost of about $60,000. The product of crystallized sugar was sold to refiners at 7 and 8 cents a pound. The yield, although not as large as expected, is still regarded as satisfactory. Owing to the "drought of unprecedented severity and length," the farmers of that region, who calculated on a yield of 10 tons of cane and 30 bushels of seed per acre, actually obtained only about 5 tons cane and 20 bushels of seed, which sold for 65 cents per bushel. This trial was so far satisfactory that it is proposed to work the product of about 1,000 acres of land for sugar the coming season (1882) at the Rio Grande Sugar Works.

October, 1882.—The foregoing statement applies to the experience of the past year. We are now able to add the following information obtained by a personal examination of the plantation and sugar works of the Rio Grande Sugar Company, Cape May County, New Jersey. This company are the present owners of the before-named works and also of 2,400 acres of land, chiefly of a light and not fertile soil, being on the peninsula between Delaware Bay and the sea, within 5 or 6 miles of Cape May and 75 miles south from Philadelphia. April 19, 1882, and following, they put in of—

	Acres.
Amber cane	958
Link's Hybrid	25
Early Orange	23
Honduras	2
By actual survey	1,008

Warned by former experience, the company determined to own and cultivate its own cane. The very cold and wet spring occasioned the loss of a considerable portion of the first planting, the loss being also due in part to deep planting by unskillful hands.*

* It is important to note here, as showing some of the practical difficulties attending the introduction of a new industry, that the farmers charged with the planting of this crop, and naturally confounding it with maize, persisted in planting the seed about 4 *inches deep*, and in rows 4 feet apart and 4 feet between hills, while in point of fact the seed requires to be lightly covered, only one inch sufficing, and in rows 3 feet apart, the plants only a few inches asunder, or in hills of many seed to the hill.

Very much of the deeply planted seed perished; while that lightly covered and in close rows made a fine "stand," requiring only to be thinned out by hand of the weak plants.

The very severe gales and torrents of rain, which swept over this county September 22 to 25, left no trace of injury in these broad acres of sorghum, as we first saw them, September 24, while the adjacent corn-fields were prostrated.

The deficient portions were replanted in June, leaving such portions of the first planting as came up, to grow together with the second planting. This circumstance worked considerably to the injury of such portions of the crop, and reduced the exponent of sugar notably. Notwithstanding this untoward circumstance, the crop, as we first saw it, near the close of September, presented a noble spectacle of vast fields of luxuriant cane ready for the rolls, and still full of vigor, and of a deep green color. The Amber cane stood about 8 to 10 feet in height. The Orange and Link's Hybrid was higher, being from 12 to 14 feet. The Amber cane only was ripe at that time, and the harvesting had been in progress from the 28th of August, at the rate of 120 to 150 tons of cane delivered daily to the mill; which is the present limit of the floors to accommodate the sugar wagons. The mill is a powerful apparatus of three rolls, each 5 feet long and 30 inches in diameter, driven by a steam engine of 125 horse-power, crushing the cane with an opening of only one-sixteenth inch between the rolls. The stalks are not stripped, only the seed heads are removed in the fields. This mill is capable of crushing 300 tons or more daily, but the floor space of the works limits the output as before stated. The product of sugar exceeds the expectations of the projectors.

The Amber cane on a large area stands not less than 10 tons to the acre on about 700 acres; the exact figures for the whole crop can be given only when the account is fully made up.* Each day's cutting is accurately recorded, and so much can now be safely stated.

We saw the "strike" of the vacuum pan of 1,600 gallons on the 28th of September, and again on the 1st of October, filling nine wagons, of one ton capacity each, with "melada" yielding $2\frac{1}{2}$ to 3 barrels of sugar to the ton.

The yield of sugar to the wagon would be, by estimate, greater by about half a barrel (the barrel holds 355 pounds) if more time could be allowed for it to stand before going to the centrifugals.

From the mill the green juice flows to a tank of 1,000 gallons capacity, whence it is pumped to defecators, after which it is hurried through the open pans to the vacuum pan, where it is reduced to about 32° B., and thence to the larger pan of 1,600 gallons, where it is raised to about 45° B., at a temperature of about 140° F. There are two "strikes" of this pan daily. The lack of space for cooling compels at present the working of the melada in the centrifugals, of which there are four, before it is completely cooled, so diminishing, as just stated, by probably a half barrel, the yield of "firsts."

We examined the books of Mr. Henry A. Hughes, the superintendent, who is a sugar-boiler of twenty years' experience, which showed the juice of the daily workings, as tested by polariscope, to have a coefficient of from 10° to 12° for the raw juice, which is polarized several times daily. For the week, ending the day of our first visit, 656 tons of cane were crushed, yielding 115 barrels of sugar of 88°, and 89 barrels of molasses of 47°. This first sugar was equal to 63 pounds to the ton of cane crushed.†

---

* About 250 acres of the land under cultivation this year were cleared of woods and shrubs too late to admit the use of lime before plowing and applying the guano. The result is very conspicuous in the diminished growth of cane, which, on this tract, is not over 5 tons to the acre, while on other land 7, 8, and 9 tons are cut on several hundred acres, and as high as 17½ tons of Amber cane, by actual survey. Ten tons of Amber may probably be a fair average product for this year, *as estimated by the superintendent at the early part of October.*

† Owing to the lack of space and the pressure of the crop, the molasses of this year's crop is held back until the crushing is over, when, unless sold at a satisfactory

The fertilizers used on the land of this plantation this year were about 25 bushels of lime, followed by 150 pounds of Peruvian guano, having as much sulphate of ammonia added as raised the nitrogen to 8 per cent. This guano cost $53 per ton. A few acres were treated, as an experiment, with fair results, with barnyard manure. On about 20 acres fish guano alone was used, the effect of which was to reduce the available sugar by about 1° on the polariscope. On the whole, the lime, guano, and stable manure gave good results. Greensand marl, which abounds in New Jersey, remains to be tested hereafter.*

The crushing of the cane with leaves settles one of the "sorghum questions" on which there has been much difference of opinion. In practice, on a large scale, the removal of leaves would involve an impracticable amount of labor. In the 1879 Report of the Department of Agriculture, p. 59, are experimental results showing an increase of both juice and sirup from the crushing of the entire plant (seeds excepted). A small loss of available sugar and a gain of sirup will probably result from crushing the blades with the stalks; a subject requiring further examination.

It is by no means improbable that in the plant's life the sucrose is elaborated directly in the leaf, and is gradually transferred to the stalk, where it accumulates.†

The full returns for the crop of this year will not be in before the closing of this report. But we are able to state, from a communication of date November 8, 1882, from the president, that the probable results of the season's work, ending with November 11, are as follows: 6,000 tons of cane; 950 barrels of first sugar, and 1,100 barrels, 50 gallons each, of molasses. The seed is not yet measured, and a full balance-sheet remains to be made up which may perhaps come in season to be added to this report.

The Orange cane turns out rather better than the Amber, being richer in juice and with an average test of 13° B.

This Committee have received from Mr. Knight, the sugar refiner in Philadelphia, a barrel of the sugar, sample of a lot of 350 barrels refined by him, of the Rio Grande Sugar Company. It ranks, on the independent judgment of experienced grocers to whom we have shown it, as "C" sugar.

---

price, it will be worked for residual sugar. It may be well to state that it is found in practice that the second crop of crystals (technically called "seconds") is about one-half the first yield, and the third crystallization gives about one-half the "second"; so, if the "firsts" are 60 pounds to the ton of cane, the "seconds" will be 30 pounds, and the "thirds" 15 pounds, or, in the aggregate, 105 pounds to the ton of cane.

The seed gathered from the Amber cane on the Rio Grande plantation, this year, measures, as we are informed by Mr. Hughes, the superintendent, 20 bushels average to the acre of cane cultivated; the yield from the Orange is 2 bushels per ton.

*The whole subject of fertilizers remains to be investigated by well-directed and carefully recorded experiments, both in the laboratory, the field, and the sugar-mill. The begasse and defecation scum pressed in cakes by the filter-press are believed to be of value as elements of fertilization; and the stubble which springs with luxuriant growth as an aftermath be of more value to the next crop if turned in as a green crop than if employed as forage. The questions of over fertility and rotation remain to be solved by experience. All we know, at present, is that the sorghum appears to thrive best for sugar on soils not too highly fertilized, and naturally of a light sandy loam.

†For additional statements respecting the Rio Grande plantation and mill, see the letters in the Appendix from Capt. R. Blakeley, of Saint Paul, Minn., and Mr. Harry McColl, a sugar planter from Donaldsonville, La. Also, in the same connection, a letter from Mr. George C. Potts, president of this company, to the Tariff Commission, and a copy of the blank form for returns of the mill, &c., required to be sworn by the superintendent to secure from the State of New Jersey the bounty provided in the act of that State, as mentioned in the letter of Mr. Potts.

Analyses of the soils of different fields are now in progress, to determine, if possible, the causes which influence such very unlike productiveness as the experience of the season of 1882 has shown to exist. The differences of yields being per acre, $3\frac{1}{2}$ (guano, no lime), $5\frac{1}{2}$ (guano, no lime), $7\frac{1}{2}$, 8, 15, and 17 tons, respectively.

3d. The Illinois Industrial University, at Champaign, Ill., have published a report " on the manufacture of sugar, sirup, and glucose from sorghum," by Henry A. Weber, Ph. D., professor of chemistry, and Melville A. Scovell, M. S., professor of agricultural chemistry (1881). The authors say:

From the approximate analysis of the [1880] cane, it appears that one acre of sorghum produces 2,559 pounds of cane-sugar. Of this amount we obtained 710 pounds in the form of good brown sugar and 265 pounds were left in the 737 pounds of molasses drained from the sugar. Hence, 62 per cent. of the total amount of sugar was lost or changed during the process of manufacture. This shows that the method of manufacture in general use is very imperfect.

In 1881 the results of an experiment on three-sixteenths of an acre of land are, as calculated on one ton of topped and stripped cane:

Pounds.
Weight of juice 834.4
Weight of sugar 77.2
Weight of molasses 119.7

In 1882 the results of the sugar-mill at Champaign, Ill., are reported as being very satisfactory to owners.

A sample of the sugar made October 19 ("product of yesterday's run of 3,600 pounds"), which reached us a few days later, was found to be of excellent quality, completely free from any trace of sorghum flavor, nearly white, and polarizing $97° .0$. We learn from the report (see Appendix, p. 78) of this year's work up to October 28 that bone-black is used at the Champaign works, in which respect they differ from the Rio Grande mill, where none no bone is used.

Professor Scovell has written us the following letter referring to the partial report, which will be found in the Appendix:

CHAMPAIGN, ILL., October 30, 1882.

DEAR SIR: I inclose a partial report of our doings at this factory this year. We will not be through grinding for a week yet, and will not be able to finish "seconds," &c., for at least a month. The report is as full as we could make it at this date. The results are creating much enthusiasm in the West. We have many visitors from abroad every day.

There is no question in my mind but that the production of sugar from sorghum will be a great industry.

Very truly,

M. A. SCOVELL.

Prof. B. SILLIMAN.

4th. The experimental farm of the University of Wisconsin, at Madison, in that State, have lately issued a report in compliance with legislative resolution, addressed to his excellency J. M. Rusk, governor, giving the results obtained in "Experiments of Amber Cane, &c., at the Experimental Farm, 1881." (See Appendix, document pp. 79–104.)

Mr. Magnus Swenson, who conducted the sorghum-sugar experiments at the Experimental Farm, has sent to this Committee a sample of both white and light-brown sugar manufactured by him from sorghum juices.

The yield of cane-sugar on two plots of land planted with Early Amber gave Mr. Swenson, for "Plot A," 923 pounds of cane-sugar and 103 gallons of sirup; for "Plot B," $907\frac{1}{2}$ pounds sugar and 87 gallons of sirup, per acre. "I separated," he says in his letter to the Chairman, "in all about 1,200 pounds of sugar, samples of which I send you." Mr. Swenson's reported details of work fully corroborate the results obtained at the Department of Agriculture. One hundred and eighty manufacturers in that State report having made about 350,000 gallons of Amber cane sirup last year, or about 2,000 gallons each.

October, 1882, Professor Swenson writes as follows of the operations of his experimental works for the current year:

UNIVERSITY OF WISCONSIN, AGRICULTURAL DEPARTMENT,
Madison, Wis., October 19, 1882.

DEAR SIR: My works on the University farm are in full operation. In spite of the very bad season the yield is very satisfactory. I am making 60 gallons of sirup per day (12 hours). This, I think, will yield at least 350 pounds of sugar and 30 gallons of molasses. The total cost of running my works, including $2.50 per ton for cane and $5 for incidentals, is $27. The analyses of the defecated juice show an average of about 10.5 sucrose and 3 of glucose. I have now ready for the centrifugal about 7,800 pounds of mush sugar.

Very respectfully,

M. SWENSON.

Prof. B. SILLIMAN,
New Haven, Conn.

Under date of October 30, Professor Swenson writes as follows:[*]

I expected when I wrote to you last to have all my "first" sugar separated by this time, but the engine I have been using was sold, and I had to rent another, which caused some delay. I have separated only 1,000 pounds. The yield has so far been 45 per cent. of the weight of the sirup. The sugar is of a light-yellow color, and is in every respect of a good quality. The cost will not exceed 5 cents per pound, even on this small experimental scale. I am very sorry that I cannot furnish you with more complete returns. I shall not be able to finish my work until the last of next week. Please remember me with a copy of your report when printed.

Very respectfully,

M. SWENSON.

5th. Capt. R. BLAKELEY, Faribault Refinery, Minnesota.—Captain Blakeley's experience in the manufacture of sugar from sorghum has already been referred to, under a former head; we add here his fuller statement of the case, which will be read with interest as a fair exposition of the reasons for a partial success which he very candidly declares to have been a commercial failure, notwithstanding a considerable output of good "C" sugar, of which he has supplied the Committee with a liberal sample.

SAINT PAUL, MINN., *April 18, 1882.*

DEAR SIR: Some time since I received your circular asking for information on the subject of manufacture of sugar from sorghum. I have been very reluctant to respond, for the reason that I did not feel that I had, as yet, much information to give, but after thinking that possibly I had as much or more than any one else upon this subject, I have felt that I ought to make known what I have done, although little has yet been accomplished.

In 1879 there was an attempt to build a sugar mill and refinery in this State at Faribault, which, like most new enterprises, wanted the proper machinery and means to make it a success, and after that season's work the company abandoned the effort.

During the winter and spring of 1880 I was induced to take hold of the property, in hopes I might be able to make the effort to make sugar from sorghum a success. During the season we made some 15,000 gallons of sirup. The cane was grown by persons who had never grown any for making sugar, and was quite inferior and badly handled; still, we made about 5,000 pounds of sugar of very nice quality, a sample of which is sent herewith. This amount of sugar was made from 1,200 gallons of sirup, or a little short of 5 pounds of first sugar from the gallon of sirup.

During the season of 1881 we made our arrangements for what we hoped would prove a more complete and successful season, but were nearly or quite as badly disappointed as the season before, as the season proved bad for the growth of cane; the harvest was very wet and the cane was nearly ruined before it could be delivered at

---

the works. Still we again made 7,000 pounds of a nice article of sugar similar to or better than the season before. But we regarded the plan upon which we had proceeded a bad one, and have suspended operations until we can place our mill upon a plantation of our own, and grow and harvest our own cane, as our experience for two seasons has convinced us that it is impossible to depend upon the farmers to properly grow and harvest the cane until they have had instruction and have been made to understand that this is the most profitable crop that can be grown in this State.

There is no doubt of the success of this industry, and if it can have the fostering care of the Government until it can be established, it will astonish the country. With good cultivation, good land, and skillful manufacturing, an acre of land should produce 1,000 pounds of sugar and 20 bushels of seed. The sugar will be worth in this market, if the present tariff is maintained, 8 cents per pound, and the seed is equal to corn to feed to cattle and hogs.

I am thoroughly convinced that this is to be one of the great industries of this country. I am sorry to be compelled to make so poor a showing, but the mill was located in a village of 10,000 inhabitants, and, under the circumstances, it was not possible to make it a success.

I shall look for the report of the Academy with the full assurance that it will confirm all that I have said.

Respectfully,

R. BLAKELEY.

Prof. BENJ. SILLIMAN,
Chairman of the Committee of
National Academy of Natural Sciences, New Haven, Conn.

6th. Mr. JOHN B. THOMS, of the Crystal Lake Refinery, Chicago, Ill., in two communications of date April 10, 1882, to Chairman imparts the results of three years' working on a large scale. He is a practical sugar refiner of eight years' experience in the West Indies. In 1879 with a "miserable mill" he obtained juice of $8\frac{1}{2}°$ B. (sp. gr. 1.060), and from a gallon of sirup weighing 11 pounds got a yield of about $4\frac{1}{2}$ pounds to the gallon. He obtained from 15 to 23 gallons of sirup to the ton of cane, weighing $11\frac{1}{2}$ pounds to the gallon, the sirup yielding $4\frac{1}{2}$ pounds sugar polarized 53° of Amber cane, which is the only sort he has worked; has known as high as 21 tons cut to an acre, and states 12 tons as an average. He sold of the crop of 1879 over 50,000 pounds of good "C" sugar, which was tested in Boston and New York, and polarized $96\frac{1}{2}$ per cent. of sugar. In 1880 his crop of about 300 acres was nearly all destroyed by a hurricane, and. the product of about 30 acres of damaged cane was all made into sirup which polarized only 42 per cent. For many details reference is made to his communications given in the Appendix p. 119. It will be observed that he cites an experiment for the production of sugar from corn-stalks (maize) in 1880, which was a failure, the stalks of sweet corn "in the milk" not furnishing juice enough to pay expenses.

7th. Mr. A. J. RUSSELL, of Janesville, Ill., writes to the Chairman, of date March 22, that he has obtained in his own practical experience "280 gallons of sirup to an acre of land, and $7\frac{1}{2}$ pounds of sugar per gallon," or 2,100 pounds sugar per acre; very light yellow, and polarized $96\frac{4}{10}$ per cent. The sirup was of a very light straw color, transparent, and free from sorgho flavor, ranking with choice New Orleans molasses. The yield of seed was from 25 to 40 bushels, and sold for 50 cents per bushel as food for stock. In an earlier communication to the Commissioner of Agriculture of date December 28, 1881, Mr. Russell states the yield per ton of cane to be from 9 to 20 gallons, and sugar from 1 to $9\frac{1}{2}$ pounds per gallon, varying with the greater or less perfection of the machinery, processes, &c. But he cites as an average in his experience, 10 tons cane per acre, 14 gallons sirup per ton of cane, $7\frac{1}{2}$ pounds of sugar per gallon of sirup, $2\frac{1}{4}$ cents per pound as cost of sugar. While, in good growing seasons on good land, he cites from experienced farmers the opinion that the product per acre is 20 tons cane, of 17 gallons sirup density per ton, and 9½ pounds sugar per gallon. If the manufacturer purchases the cane from the farmer, the cost of the sugar is put at 3¼ cents per pound. These letters are in the Appendix, p. 118.

8th. GEORGE W. CHAPMAN, secretary Rice County Farmers' Club, Kansas, writes of date February 4, 1882, to the honorable Commissioner of Agriculture:

> I worked up last season 75 acres of cane, Amber and Honduras. Amber yielded 9 tons stripped and topped, and the Honduras 33¼ tons raw stalk to the acre, being the largest yield of cane yet known in Kansas. * * * I made some sirup by an evaporator and it all granulated to a solid.

From such a yield of cane as the Honduras here named it is easy to obtain over 3,000 pounds of crystallized sugar to the acre and 100 gallons of sirup. (See his letter in Appendix.)

9th. While the sorghum sugar industry is comparatively an unexplored field, no attempt having been made heretofore in any European country to investigate systematically the conditions requisite for successful cultivation and utilization of the sorghum cane, the necessity of such investigations is strongly insisted upon by M. Basset, one of the highest French authorities on the sugar industry, who devotes some 32 pages of the third volume of his "Guide Pratique des Fabricant de Sucre" (Paris, 1875) to an impartial discussion of the question of the extraction of sugar from sorghum and maize stalks. On page 217 of the volume in question Basset describes two experiments on canes grown by himself at Paris from which he extracted in the first case 8 per cent., in the second 11 per cent., of the weight of the canes in crystallized sugar. In the first experiment the defecated and concentrated juice granulated after the first boiling in *four* days; the second product from the molasses of the first product took a longer time for crystallization. On page 218 the same author expresses himself as follows:

> Nous croyons donc que le sorgho offre un avenir sérieux à l'industrie sucrière, et que cette plante, susceptible d'entrer avec avantage dans nos assolements culturaux, est appellée à devenir en France un puissant auxiliaire de la betterave. Il ne faut pas songer à opposer ces deux végétaux l'un à l'autre; ce serait un acte de véritable folie; la betterave est indispensable à notre agriculture, et la prospérité de cette plante est liée à celle de la production du pain, de la viande, de la laine, etc.; ainsi qu'à l'amélioration du sol, on ne peut songer à supplanter ou à remplacer cette précieuse racine. Cependant, pourquoi rejetterait-on sans examen sérieux et sans expérimentation, par mauvais vouloir et parti pris, un végétal dont la richesse saccharine peut *venir en aide* à la production sucrière? Ce serait evidemment une faute impardonnable, et la fabrication comprendra facilement que c'est à elle à prendre les devants et à entrer dans une voie d'amélioration que la culture négligera pendant longtemps si elle ne se sent pas soutenue par l'industrie.

10th. Mr. S. W. JOHNSON, a member of this Committee, states that Messrs. Doolittle and Bartlett, farmers in New Haven County, Connecticut, have for many years made a successful business on a considerable scale in growing sorghum and making melada for the supply of their neighbors in a home market. From this melada Mr. Johnson has prepared the two samples of "C" sugar submitted to the Academy herewith (samples of melada marked X and Y, and of crystallized sugar XX and YY).

These results are obtained by open-pan evaporation and without special skill. The only point in which they differ from the practice of those who have produced chiefly sirup (glucose) without sugar is that they have permitted the canes to mature. The sirup made by them weighs 11 to 12 pounds per gallon, and crystallizes, on standing, into melada.

11th. CLINTON BOZARTH, Cedar Rapids, Iowa, in his address before the Cane Growers' Association at Saint Louis, in January, 1882, as quoted in the Proceedings of that Convention, p. 19, gives in brief his successful experiments in the production of sugar and sirup on a large scale for twenty years. Samples of Mr. Bozarth's melada, sirup, and crystallized cane sugar are before us.

12th. *Proceedings of the Mississippi Valley Cane Growers' Association.* 8vo., pp. 36. Saint Louis, 1882. This report contains the address of Mr. Bozarth, the above named, and numerous data from various cultivators, together with a carefully considered address of Prof. W. H. Wiley, of La Fayette, Ind., on adulterations.

13th. *Sugar-Canes, and their products, culture, and manufacture.* By ISSAC A. HEDGES. 12mo., pp. 190, 1881. In this revised and enlarged edition of his well-known book, Mr. Hedges, who is a veteran in sorghum-sugar production, has brought together a considerable amount of important matter and original data bearing on this subject, which the Committee have consulted with advantage, and to which reference is made for many details of interest in the history of the development of this industry.

14th. JOSEPH S. LOVERING (1857), "*Sorghum saccharanum,* or Chinese Sugar-Cane."—Mr. Lovering's original memoir has become rare. It is, however, reproduced in Mr. Isaac A. Hedges's volume on "Sugar-Canes," pp. 123–140. The experiments of Mr. Lovering are of especial interest as showing how early in this industry many of the important points needful for success in producing sugar from sorghum were clearly recognized and laid down. Mr. Clinton Bozarth, of Cedar Falls, Iowa, whose sugar samples are before us, says, in his late address at Saint Louis, that he has for over twenty years followed the rules laid down by Mr. Lovering with success. Yet so slowly do the most clearly stated principles reproduce themselves in practice that comparatively few cultivators have followed the example of Mr. Bozarth.

15th. *The Jefferson Sugar Manufacturing Company, manufacturers of sugar and sirups from the juice of Cane and Corn Stalks.*—The following letter from Mr. Henry Talcott, president of the company named, was written after his visit to the Rio Grande Sugar Works, and is of special interest from his statements respecting the absence of any ill effects of frost upon his cane, after repeated sharp freezings. His success with open-pan boiling is also valuable to the small farmer, and his final report after "swinging out" the sugar of his crop will be of general interest.

JEFFERSON, ASHTABULA COUNTY, OHIO, *November* 2, 1882.

DR. PETER COLLIER,
Chemist, Department of Agriculture, Washington, D. C.:

* * * I have been endeavoring to secure a practical method of producing the same results which they have obtained at the Rio Grande Company's works—where I have just been to see for myself—that our farmers could all adopt with small means, and make this industry universal. I think our company can show the world as complete success in about four weeks as the Rio Grande have done, on a much smaller and simpler scale. We are now crushing and boiling from 10 to 15 tons of cane stalks daily; have been doing this for four weeks past. Our returns in yield are the same in substance as the Rio Grande, but, unlike them, we have had ten or fifteen good hard, white frosts, some of them hard enough to freeze ice on water thick as window-glass. Our cane was standing in the fields; we are yet cutting it. I had 10 acres of it on my own farm. We see no ill effects from it in our work; we have made just as good a yield of juice; it makes just as good sirup and sugar; and all we have lost, as far as we can discover, is the leaves for our cattle fodder. Mr. G. C. Potts wished me to notify you of this fact on my return home, also to send you some samples of our work. We cook in open pans, by the Stewart process, only much more perfect than

he ever did his work (except in theory). F. C. Knight analyzed our mush sugar and finished sugar yesterday in their refinery, and pronounced it the purest and best sugar they ever saw. The sugar was our "second." This year's stock is still in our hot room granulating slowly, for we dare not cook it dry in open pans, for we are so liable to scorch it when near done; so we make time and warm room do part of the work. We shall not use our centrifugal until the close of this month; shall then have from 60,000 to 80,000 pounds of mush to work over. I shall make as complete and clear report of it to the Department as I possibly can; I shall also visit the Champaign Works in Illinois next week, and compare notes with them. I have an invitation to do so, and must see the bottom of this industry so far as it is practically developed. Of course the vacuum pan and animal-bone filter make the refined sugar at once—a specimen of it they sent me yesterday, and I inclose a little of it for you—but this expensive machinery, if it is more profitable, cannot be made to come in general use. Our farmers must do this work as handy as they can make good butter or cheese, to get them into it in any great numbers. Our factories are learning, many of them, to do the work, and several others are to-day making mush sugar at their own molasses factories, while we furnish them solution B and do their centrifugal work. I will send a little sample of sugar we purged yesterday for Mr. P. A. Upp, of Edgerton, Williams County, Ohio, who made it under our directions, and then brought to the factory to see our works, and with his own eyes see finished sugar of his own make. I guess he was as well pleased with the result as any fond mother could well be with her first-born. He returned home with his sugar, and said he should now go shouting among his own people, for he had accomplished well what his people all said was an impossibility.

Yours, respectfully,

HENRY TALCOTT.

16th. B. V. RANSOM, of Salem, Nebr., is one among many farmers who has been successful in making both sirup and sugar from sorghum in a small way, and especially he deserves respectful notice for his accurate statements of his observations for 1881 and 1882, the full details of which will be found in the Appendix (p. 125). It is interesting to observe his remarks on the varieties of sorghum which he cultivates, and his mode of manufacture with open boiling and time defecation, by which he has worked out his success in a plain common-sense way.

17th. C. CONRAD JOHNSON, Baltimore, Md., a sugar master of practical experience in San Domingo, W. I., has submitted to the Committee an elaborate statement expressing his views on the subject of sorghums, with reference to the prospective production of sugar from their juices as appears from an examination of Dr. Collier's results. Mr. Johnson's communication appears to the Committee so valuable, in view of his familiarity with the whole subject of sugar production, and the very practical view he takes of the investigation, and of the probable future of this industry, that his letter is presented in full in the Appendix, page 131.

PART II.

CONCLUSION AND SUMMARY.

CONCLUSION.

The Committee find, as the result of their investigation, by all the data which have come before them, as well as those obtained by the Department of Agriculture during the years from 1878 to 1882, both inclusive, and those derived from other parties in different sections of the United States, that the following points are established by an amount of investigation in the laboratory, and of practical experience in the field and factory, which have rarely been devoted to the solution of any industrial problem.

The more important and well-established results are here enumerated, and are followed by a statement of certain practical and scientific points which still remain for future inquiry.

A.—OF THE POINTS ALREADY SETTLED.

1.— THE PRESENCE OF SUGAR IN THE JUICES OF SORGHUM AND MAIZE STALKS.

From records examined by this Committee, it appears that, during the three years prior to 1882, there have been made at the Department of Agriculture almost four thousand five hundred chemical analyses of the juices of about forty varieties of sorghum and of twelve varieties of maize. These analyses have shown the constitution of the juices of each variety at the successive stages in the development of the growing plant. They not only confirm the well-known fact of the presence of sugar in the juices of these plants in notable quantity, but they also establish beyond cavil, what seems surprising to those who have not examined the facts, that the sorghum particularly, holds in its juices, when taken at the proper stage of development, about as much cane-sugar as the best sugar-cane of tropical regions.

An examination of the analytical tables in the reports of Dr. Collier, synopses of which follow, will show that the juices of sorghum in certain exceptional, but not isolated, cases were remarkable for the amount of cane-sugar they contained, viz:

Of true crystallizable sugar in the juice—

	Per cent.
5 analyses of five varieties gave over	19
3 analyses of 17 varieties gave over	18
79 analyses of 23 varieties gave over	17
152 analyses of 30 varieties gave over	16

* Even at the risk of repeating some statements already made in the earlier portions of this report, the Committee consider it is better to review systematically in this summary the whole ground they have gone over.

43

As compared with the juices of sugar-cane, which gave by analysis under 15 per cent. of sugar, these results are unexpected and surprising.

But the average results obtained during long periods of working and from different varieties are of more value to the practical farmer than any exceptional instances.

The average results obtained from 122 analyses of 35 different varieties of sorghum, and during a working period of one or another of the above varieties of at least three months in the latitude of Washington, are as follows:

Average results of analyses of juices of 35 varieties of sorghum.

	1.	2.	3.	Average.
Sucrose...................................per cent..	15. 99	15. 94	16. 61	16. 18
Glucose.................................do....	1. 84	1. 72	1. 83	1. 80
Solids ....do....	3. 01	3. 20	3. 01	3. 08
Available sugar..........................do....	11. 14	11. 02	11. 77	11. 30
Juice.................... ....do....	60. 25	58. 95	56. 51	58. 57
Specific gravity of juice	1. 082	1. 081	1. 081	1. 0813
Number of analyses	40	37	45	122

From this statement it will be seen that, as an average of all the analyses made during those three stages, there was obtained 58.57 per cent. of the weight of the stripped stalks in juice; that 16.18 per cent. of the weight of this juice was crystallizable cane-sugar; and that 11.30 per cent. of the weight of the juice may be obtained as sugar by the ordinary process of manufacture.*

By reference to the tables it will also be seen that of the eight varieties of maize examined in 1881, seven of which were of common field and one of sweet corn—

Per cent. of cane-sugar.

3 analyses of 3 varieties gave over....................................	13
9 analyses of 7 varieties gave over.......................................	12
22 analyses of 7 varieties gave over......................................	11
29 analyses of 7 varieties gavé over......................................	10
35 analyses of 7 varieties gave over......................................	9

Of ten varieties of maize grown in 1880, the following results were obtained :

Per cent. of cane-sugar.

124 analyses of 10 varieties gave over....................................	9
90 analyses of 10 varieties gave over.....................................	10
59 analyses of 9 varieties gave over......................................	11
24 analyses of 9 varieties gave over......................................	12
8 analyses of 4 varieties gave over.......................................	13
2 analyses of 1 variety gave over	14
1 analysis of 1 variety gave over	15

In 1880 over sixty-two millions acres of our land were in maize, or 38 per cent. of all the cultivated land of the United States. The amount of sugar thus apparently lost, calculated on the results obtained at the Department of Agriculture in the last three years, is equal to the present product of the entire world. It is premature to say that the profit-

* The " available sugar" here stated is the amount of cane-sugar shown by analysis, less the sum of the glucose and solids not sugar; e. g., in this case 16.18 per cent. less 1.80 per cent. + 3.08 per cent. = 11.30 per cent. This mode of computation as has already been explained, gives a less probable quantity of available sugar than is shown by the method of " exponent," usually used by sugar-boilers.

able extraction of sugar from corn-stalks is demonstrated, but such a result may yet be possible.*

2.—PRACTICALLY LITTLE DIFFERENCE IN THE VARIETIES OF SORGHUM AS TO THEIR CONTENT OF SUGAR.

The results of the investigations at the Department of Agriculture have shown the remarkable uniformity of the several varieties of sorghum as sugar-producing plants when fully developed; and have also shown the different varieties to vary widely in the time required for their full development, varying, as has been shown, year after year fully three months as between the earlier and later maturing varieties.

This fact of the wide variation in the different varieties in their period of reaching full maturity, although previously recognized, has not received the consideration which its extreme importance demanded, as is evinced by the fact that at present, as for the past thirty years, those varieties are largely grown in the Northern States which could only reach maturity at rare intervals and in exceptional seasons in these latitudes. This satisfactorily accounts for the occasional production of crystallizable sirups, and the general failure to secure similar results continuously.

3.—WHEN THE MAXIMUM CONTENT OF SUGAR IS PRESENT IN THE SORGHUM.

No conclusion established by the work of the Department of Agriculture, practically considered, is of greater importance than the positive ascertainment of that period in the development of the several varieties of sorghum when their juices contain the maximum of cane-sugar.

4.—CONFLICTING TESTIMONY BEFORE THIS INVESTIGATION.

On this point there has existed, during the past twenty years or more, the greatest discrepancy in statement, and the general opinion prevailing has been very wide of the truth, as established by all these experiments.

As evidence of the great diversity of opinion concerning this important matter which existed previous to the experiments at Washington, the following quotations are made from the reports of various experimenters:

a. In his report on "Early Amber Cane," by Dr. C. A. Goessmann, of Amherst, Mass., 1879, he says, p. 9:

The safest way to secure the full benefit of the Early Amber Cane crop for sirup and sugar manufacture is to begin cutting the canes when the seed is full grown, yet still soft.†

---

* The only trial on a large scale for extracting sugar from corn-stalks of which we have record will be found in the statement of J. B. Thoms, of date April 10, appended to this Report (p. 119), and was not a success. It is possible that if the maize had been allowed to mature, in place of being cut when the ear was in an immature state fit for canning, the result might have been different.

† Dr. Goessmann's statement requires the modification as explained by him in Committee viz: To secure the crop (some 20 acres of land), it was essential, with the limited milling power at his command, to commence milling at the time specified, even if some loss of cane-sugar followed this course.

b. In the "Sorgo Hand Book," published by the Blymyer Manufacturing Company, Cincinnati, Ohio, 1880, it is directed upon page 8:

The cane should be cut when the seed is *in the dough*, and *several days before* grinding, as it will be more free from impurities if cured for a few days before going to the mill.

c. In a pamphlet entitled "Sugar-making from Sorghum," published by the Clough Refining Company, p. 5, directions are given to—

Harvest as soon as the *seeds begin to form*, and *before they get hard*. *Grind the cane, if possible, soon after it is cut.*

d. In a pamphlet entitled "The Sorgo Manufacturers' Manual," by Jacobs Brothers, Columbus, Ohio, p. 4, 1866, it is stated that—

The cane is in the best state for harvesting when part of the seed is beginning to turn black, or in other words, *when the seed is in a doughy state*. The cane should be cut and shocked in the field with tops on, and in this condition it may remain several months before being worked up, for the cane matures and forms more saccharine matter.

e. In a "Report on the Manufacture of Sugar, Sirup, and Glucose from Sorghum," by Professors Weber and Scovell, of the Illinois Industrial University, 1881, p. 22, they say :

The proper time to begin cutting the cane for making sugar is *when the seed is in the hardening dough*. The cane should be worked up as soon as possible after cutting.

f. J. Stanton Gould, in a "Report on Sorghum Culture," made to the New York State Agricultural Society in 1863, p. 752, says :

The seed of the cane (sorghum) continues in the dough for about a week. It is the general impression the cane should be cut during this period, as it is then *supposed* to have the greatest amount of saccharine matter; at least, this is *thought* to be true of all the varieties except the White Imphee, which is usually cut *just as it is going out of the milk or just entering the dough.* •

g. In conclusion, we quote from Mr. Gould's paper, as illustrating the chaotic state in which our knowledge was prior to the work at the Department of Agriculture. Upon p. 740 he says :

These conflicting opinions might easily be reconciled by a few well-directed experiments.

Again, he says, same page:

After the most careful inquiry, orally and by letter, I am unable to find that any such experiments have ever been made.

Again, he says, p. 747 :

These experiments are not conclusive, and the whole question needs a careful and accurate investigation.

As the result of such an investigation, we call attention to the average results of the past years, as shown in the tables accompanying this report, from which it will be seen that during each of the past three years it has been demonstrated beyond any reasonable doubt that the value of the sorghum for the production of sugar increased, upon an average of the 35 or 37 varieties tested, fully 500 per cent., and in many cases 1,000 per cent., after the period when, according to the authorities cited, it was recommended that the crop should be cut up.*

---

* MAY, 1883.—The references in the text to "the tables accompanying this report" were made originally upon the expectation that the full text of the documents referred to the Academy by the Commissioner of Agriculture would be reproduced as a part of this Report, as well, also, as the "*Graphical Charts*," essential to a full understanding of the results upon which this Report is based. The restrictions imposed by the terms of the Senate's resolution of March 3, 1883, seriously impairs the value of this Report as a Sorghum Manual, by suppressing the documents in question, with their illustrations. Of the Department's Report for 1882, an edition of 300,000 copies has

It will be observed also how completely at variance the above quoted authorities are in reference to the subsequent treatment of the crop after cutting it up, the one recommending that it be stored, even for months; the other, that it be immediately worked up. The importance of this latter course of treatment can hardly be overestimated, as appears. from data herewith presented.

5.—THE IMPORTANCE OF AN EVEN CROP, WITH NO SUCKERS, IN THE
\ PRODUCTION OF SUGAR.

The experiments at the Department of Agriculture this past season have fully confirmed the practical wisdom of a course which is pursued by the sugar planters of Louisiana and Cuba, viz, the exclusion from the matured crop of all immature canes, if the production of sugar is contemplated.

This point, if previously recognized by sorghum growers; has never been properly understood and considered as it deserves to be.
\

6.—THE IMPORTANCE OF PROMPTLY WORKING THE CROP AFTER IT
HAS BEEN CUT UP.

To this point also reference has been made already. Its importance can hardly be overstated. If departure from this rule is *at any time* admissible, it is at least safe to say that the conditions which would warrant such departure are as yet not determined. Prompt working of the cane so soon as cut is always safe, and any delay is fraught with unavoidable risk of loss.

This conclusion is established, as well by the work of Dr. Goessmann as by that of the Department of Agriculture.

7.—SUGAR HAS BEEN MADE FROM SORGHUM AND CORN-STALKS.

It will be seen from the reports of the past three years at the Department of Agriculture, as well as from a wide range of experience elsewhere, that sugar in large quantities has been shown to be present in the juices of sorghum and maize. Also, which is of the first importance from the economical side, high-grade marketable sugar in con-

---

been printed. Of the earlier reports, containing Dr. Collier's results on sorghum, it is understood the editions are exhausted, including Special Report No. 33.—[COMMITTEE.]

The Senate's resolution, above referred to, is as follows:

IN THE SENATE OF THE UNITED STATES,
March 2, 1883.

Resolved by the Senate (the House of Representatives concurring), That the Report of the National Academy of Sciences on the sorghum sugar industry be printed with such portions of the appendix and accompanying exhibits as may be selected by the Joint Committee on Public Printing, and that there be printed 6,500 additional copies, of which 2,000 copies shall be for the use of the Senate, 3,000 copies for the use of the House of Representatives, 1,000 copies for the use of the Department of Agriculture, and 500 copies for the use of said National Academy of Sciences.

Attest:
F. E. SHOBER, *Acting Secretary.*

IN THE HOUSE OF REPRESENTATIVES,
March 3, 1883.

Resolved, That the House concur in the above resolution of the Senate.
Attest:
ED. MCPHERSON, *Clerk.*

siderable quantity has been successfully made at various places, as already cited, from sorghum juice, comparing favorably with the sugar from the true sugar-cane or from the sugar-beet.

The testimony of the sugar boiler at the Department of Agriculture who worked up the sorghum in 1881, and who produced a sugar which polarized 97.5 per cent., is of especial value. He says in his report that "sugar of this character could have been produced day after day from sorghum such as produced this"; and also, in reference to this sorghum, he testifies " it was only fairly good."—(*Vide* report 1881-'82, Peter Lynch.)

It will be seen that in successive years there was also obtained from the stalks of common maize, *after the ripened grain had been plucked*, at the rate of 900 pounds of sugar to the acre. It also appears from the correspondence submitted that many parties have practically secured results nearly equal to these in their work.

8.—THE HYDROMETER AND RIPE SEED SUFFICIENT TO INDICATE THE PROPER TIME FOR WORKING UP THE CROP.

It will be seen by reference to the reports of the work at Washington that it is within the means of the common farmer to inform himself accurately as to the condition of his crop by simply examining the seed, and by the hydrometer learning the specific gravity of the expressed juice.*

By reference to the preceding reports of the Department of Agriculture it will be seen that for each increase of .001 in specific gravity between 1048 and 1086 in the year 1880 there was an average increase (glucose excepted) in the several constituents of the juice of the several sorghums as follows :

Per cent.

Sucrose .... .251
Solids .... .067
Available sugar .... .257
Glucose ....minus.. .073

Number of analyses, 2,186.

In 1881 the increase for each .001 specific gravity was, in the average results, as follows, for specific gravity between 1052 and 1082 :

Per cent.

Sucrose .... .305
Solids .... .013
Available sugar .... .354
Glucose ....minus.. .062

Number of analyses, 438.

The general average for the years 1879, 1880, and 1881, specific gravity between 1048 and 1080, was as follows for each increase of .001 specific gravity :

Per cent.

Sucrose .... .238
Solids .... .028
Available sugar .... .262
Glucose ....minus.. .052

Number of analyses, 2,960.

---

* The Committee do not wish to be understood as advising every farmer to be his own sugar boiler. While it is probably quite true that with very simple means and moderate skill good crystallizable sirup may be made on the farm, it is clear that the skill and experience of a professional sugar-master is essential to the successful management of the trains and vacuum-pans of a well-ordered sugar-house, and the natural result will be, beyond doubt, that such establishments will be set up at convenient points in each sugar-producing district. The problem, as far as it relates to the production of sirups, appears to be already solved by abundant experience.

For changes in specific gravity in successive stages of development, each increase of .001 specific gravity corresponded to the following results:

Specific gravity.	Sucrose.	Solids.	Available sugar.	Glucose.	Number of analyses.
	Per cent.	*Per cent.*	*Per cent.*	*Per cent.*	
1018 to 1029	.066	.016	—.034	.084	146
1029 to 1042	.122	.025	.069	.028	191
1042 to 1052	.290	.011	.062	.017	129
1052 to 1061	.299	.010	.340	—.051	158
1061 o 1071	.273	.023	.395	—.055	137
1071 o 1082	.317	.011	.371	—.065	236

From these it will appear that the sorghum juices, after they have reached a specific gravity of about 1050, increase gradually and with great regularity in saccharine strength and in available sugar until a specific gravity of 1080 to 1082 is attained, and that this increase is fully, upon the average, 0.3 per cent. of the weight of the juice for each .001 increase in specific gravity, or an average increase between 1050 and 1082 of 9.6 per cent. of the weight of the juice in available sugar.

The practical importance of this fact, which appears to be demonstrated by the very numerous analyses made during the past three years, can hardly be too strongly emphasized.

By reference, then, to the table given upon page 79, Special Report 33, the farmer may, by simply taking the specific gravity of his sorghum juice, readily estimate the approximate value of the crop for the production of sugar or sirup.

9.—LENGTH OF PERIOD FOR WORKING SORGHUMS.

Reference has already been made to the very great difference existing between the different varieties of sorghum as to the length of time needed for them to reach maturity. It is not known that experiments have been made to determine this difference accurately, until those lately made at the Department of Agriculture. It has also been shown, as already remarked, that those varieties requiring long periods for their complete maturity have been the varieties largely cultivated in the Northern States during the past thirty years.

The results given in the Special Report No. 33, page 96, Table 96, show not only the number of days from time of planting to complete maturity of each variety, but also the number of days during which the several varieties were in a condition for working in this latitude.

By this table the farmer in any section of the country may be able to select such varieties as the nature of his climate will give him reason to believe may be successfully grown; or, if his season permits, he may select several varieties, which, coming to maturity in succession, will enable him to extend his working season, and yet have his cane of each sort in the best condition for sugar or sirup production. Planted, as these several varieties were, side by side in the same soil and on the same day, the comparative results given in the table referred to are fully trustworthy, and could have been secured in no other way.

These results are of direct practical value to the sorghum grower, and were confirmed by the experience of the past season.

10.—EFFECT OF RAIN UPON THE COMPOSITION OF SORGHUM JUICES.

The investigation of this question and the results secured offer a good illustration as to the importance of submitting doubtful questions to the test of actual experiment, since it is nearly certain that any one, reasoning from *a priori* considerations, would have concluded, and indeed such conclusion has been accepted as established fact, that the effect of rain would be manifest in a diluted juice, and that conversely a prolonged drought would result in a concentration and diminution of the juice. The results, however, of very many experiments on every variety of sorghum, during the past season, prove the incorrectness of such conclusions.

11.—THE EFFECT OF FROST UPON SORGHUM.

The investigations concerning this question practically reconcile the discordant reports in regard to this matter. It has been shown that when fully matured the sorghum withstands even hard frosts without detriment, but that if immature the effect is most disastrous.

It is shown also that this disastrous result is due not directly to the effect of the frost, but to the subsequent warm weather, which rapidly induces fermentation with inversion of sugar in the frosted and immature cane.

12.—MANUFACTURE OF SUGAR FROM SORGHUM.

From the numerous results given in Dr. Collier's reports, it is obvious that the method of manufacture of sirup was such that nearly all of the sugar present in the juices of the sorghum or maize could be secured in the sirup without inversion. This point is one of especial importance practically, and since the results differ so widely from those of other experimenters, they are entitled to careful consideration.

A single experiment of Dr. Goessmann gave from a juice containing 8.16 per cent. sucrose and 3.61 per cent. glucose a sirup containing 37.48 per cent. sucrose and 37.87 per cent. glucose, or as follows:

	Per cent.
Juice:	
Sucrose	69. 33
Glucose	30. 67
Sirup:	
Sucrose	47. 94
Glucose	52. 06

From which it appears that, supposing there was no loss of glucose in the operation of making the sirup, 21.39 per cent. of the sucrose was converted into glucose, or, in other words, 30.85 per cent. of the sucrose in the juice was inverted. If such a result was to follow invariably, no one, we think, would hesitate to accept the following conclusion drawn by Dr. Goessmann from the above experiment, viz:

In sight of these facts it will be quite generally conceded that the sugar production from sirup like the above must remain a mere incidental feature in the Amber-Cane industry in our section of the country.

In 1879 the average of 24 experiments with the juices of several varieties of sorghum and maize, made at the Department of Agriculture (see Annual Report 1879, p. 53), showed that the relative loss of sucrose in the sirup was only 5.35 per cent. of that present in the juice, instead of being, as Dr. Goessmann found, 30.85 per cent.

But of far greater importance is the fact brought out in an average of 40 experiments, including all made, that there was an actual loss of only 12.5 per cent. of the cane sugar; *i. e.*, there was secured as sugar in the sirup 87.5 per cent. of all the sugar present in the juice; thus showing that even the total loss by defecation, by skimming, and by inversion, was no more than that usual with sugar-cane juice, for it is estimated that only about 80 per cent. of the cane sugar present in the tropical juices is recovered in the sugar and molasses, a little over 20 per cent. being lost in the manufacture.

In Ure's Dictionary, Appleton's edition, 1865, vol. II, p. 758, the writer upon sugar says as follows:

The average quantity of grained sugar obtained from cane juice in our colonial plantations* is probably not more than one-third of the quantity of crystalline sugar in the juice which they boil.

13.—EFFECT OF FERTILIZERS UPON THE PRODUCTION OF SUGAR IN SORGHUM.

Many experiments have been made with a view to the determination of this question. A limited number of these being taken, conclusions apparently well established would follow from such limited examination. The result, however, of all the experiments, including 34 analyses of the ash of juices from sorghum grown upon plats differently fertilized, leaves the matter wholly undecided.

In the literature of sorghum respecting fertilizers very much may be found, as in that upon sugar cane and beets, which appears to be well established, at least it is with great confidence asserted; but it is very doubtful whether any conclusion as to the effect of one or another fertilizer upon a sugar-producing crop rests upon data involving over 34 analyses of ash, and, as has been said, even this number fails to prove anything as to the effect of various fertilizers upon sorghum.

14.—THE SO-CALLED GUM A PRODUCT OF MANUFACTURE.

In the purging of sorghum and corn-stalk sugar, it happens very often that this operation is of unusual difficulty, owing to the presence of a certain gummy substance, and this practical difficulty has been by some so magnified that the economical production of sugar from these two plants has been confidently declared impossible.

In this experience in Washington, as well as that of many other observers, this peculiar substance has been found often to be present in quantity so small as to offer little, if any, resistance to complete purging in the ordinary centrifugal.

It is a matter of very great practical importance to determine those conditions which prevent its being produced in the manufacture of the sirup, since in no case has its presence been detected in the freshly expressed juices of either sorghum or maize. It appears to be formed by transformation of other constituents of the juice in *the process of sirup production.*

B.—FUTURE INVESTIGATION.

Although much important work has been already accomplished, and results fully repaying the care and expense bestowed have been attained, there yet remains a vast amount of work demanding further investigation. Even granting that the questions already settled may suffice to place this new industry upon a safe and profitable footing, it by no means follows that it may not be made more profitable.

Under the careful supervision of science from its earliest infancy, the beet-sugar industry has so advanced that to-day 38 per cent. of the world's supply of sugar is derived from this source—a plant poorer in sugar, more expensive in cultivation, and far more difficult and costly in the means required for the extraction of its sugar than sugar cane—and yet under this scientific supervision it stands practically the sole rival of the cane as a source of supply for sugar.

---

* British possessions.

Perfected processes and improved appliances have enabled the manufacturers to obtain practically all of the sugar present in the beet, either as sugar or molasses or spirits, while, on the other hand, it is estimated that fully one-third of all the sugar in the cane is burned up in the begasse upon the sugar plantation.

The same methods, the same apparatus, the same waste which are in use and characterize our production of sugar from cane obtains in its production from sorghum. Sixty per cent. of juice from an actual 90 per cent. is the maximum yield of our cane mills. This, then, remains a matter for future experiments and solution.

The effect of fertilizers upon the growth of the sorghum and of maize and upon the composition of their several juices yet remains in a state of entire uncertainty.*

The variety of soil best adapted for the production of sugar in these plants is equally a matter of which we are in comparative ignorance. We have at present six varieties of sorghum which for centuries have been grown in Northern China, and thirteen varieties sent from Natal. It is well known that these countries are the sources whence we originally obtained our varieties some thirty years ago. It would be a most interesting question to determine whether these seeds direct from China and South Africa would grow canes as rich in sugar as those already examined, or whether in our climate and soil this sugar-producing quality has been developed during the past thirty years in which they have been grown here.

Should this latter prove to be the case, it would give us reason to hope that improvements equally great might possibly follow; or that by careful crossing or selection a variety could be secured surpassing as a sugar-producing plant any of the numerous varieties now known.

The methods of defecation in the process of manufacture are completely unsettled, and the greatest difference of opinion and practice prevails among cultivators and manufacturers.

The use of lime or of some other alkaline agent, the removal of the sediment, and the treatment of both the scum and the precipitate demand further investigation.

The same is true of the use of sulphurous acid or oxide, in solution or in vapor, which is open to many doubts in the minds of sugar-masters—doubts which may be empirical, but which careful research alone can dispel or confirm. It is worse than idle to dogmatize on matters of this description, but dogmas will prevail where sound evidence is wanting.

There are chemical agents which may be tried in connection with sorghum sugar production of which as yet we have no recorded experience and no laboratory guidance; for example, the action of sulphites and hyposulphites of the alkalies and of alkaline earths in place of sulphur fumes or sulphurous acid.

There is a wide range of experiment possible in the methods of clarification by other agents than those familiar at present.

We are ignorant of the possibilities which may attend the attempt to reduce the sucrose to an insoluble lime salt which can be kept indefinitely and transported as flour.

The extensive literature of the sugar industry, enriched during the

---

*The results of Prof. Magnus Swenson, of Wisconsin, (Appendix, p. 146,) are the latest which have been communicated to this Committee (November, 1882) on the use of fertilizers, and these appear to show the inefficiency of nitrogenous manures on the sugar production of sorghum. This is also the conclusion from the use of guano at the Rio Grande plantation, where it had a fair trial in 1882. The whole subject of fertilizers require more extended research on a systematic plan, with varied soils.

half century or more since the days of Napoleon I by the labors of the best technical chemists of Europe, is far from being exhausted in the search for data long slumbering in almost forgotten pages from which important suggestions may arise in aid of the sorghum industry.

We must not rest until an economical and rapid method is discovered to save the loss of about 40 per cent. of the juice which is now wasted in the begasse. Such an invention would enrich the world equally in the tropics and all cane-growing countries as in the fields of sorghum. But such methods are perfected only as the fruit of research, and this must not be relaxed when on the verge of success.

The Committee have not taken up the fodder question in connection with this general discussion. It did not properly fall within the range of nquiry assigned to them by the Academy. But it is conspicuous that it is a subject of great moment germane to this investigation, inasmuch as t is closely related to the best use of " waste material," and yet more so if we consider the surprising fecundity of the sorghum stubbles, whether it is grown for green fodder or for soiling. It will be found on consulting the records of the Department of Agriculture that a notable amount of good work has been done in this direction by the Chemical Division, and it is clearly desirable that it be made a subject of further inquiry.

The spirit of scientific investigation which has led the Department of Agriculture through its chemical and agronomic researches to results of such importance towards developing a new industry of national value has been liberally fostered by the General Government, and to some extent also by certain of the States. The fruits of this policy are already beginning to show themselves in the decided success which has attended the production of sugar from sorghum on a commercial scale in the few cases in which the rules of good practice, evolved especially by the researches made at the laboratory of the Department of Agriculture, have been intelligently followed. Sufficiently full returns from the crop of 1882 have already come to hand to convince us that the Industry is probably destined to be a commercial success.

The practicability of separating sugar from sorghum has been abundantly shown in a multitude of examples. But the Committee are of opinion that many important practical questions, yet unsettled—some of which have been indicated in this report—can be better solved by the means of research now to be found in public institutions, and more especially in the laboratories and experimental works of the Department of Agriculture than elsewhere; and that the sugar-producing industry of the whole country, both that of the tropical cane in the South and the sorghum over a far wider area, will derive yet greater benefits from the continued investigations of the chemist of this Department, to whose former work we are already so much indebted.

B. SILLIMAN, M. D., &c.,
Professor of Chemistry, Yale College, Chairman.
WM. H. BREWER, Ph. D.,
Norton Professor of Agriculture in Yale College.
C. F. CHANDLER, Ph. D.,
Professor of Chemistry, Columbia College, New York.
S. W. JOHNSON,
Professor of Agricultural Chemistry, Yale College,
Director of the Connecticut Agricultural Experiment Station.
GIDEON E. MOORE, Ph. D.,
New York.
J. LAWRENCE SMITH, M. D.,
Louisville, Ky.

PART III.

APPENDED PAPERS.

1.—OF THE SO-CALLED "CHINESE SUGAR-CANE."

Dr. S. WELLS WILLIAMS, the learned Sinologue, whose remarkable familiarity with Chinese literature and natural history entitles his statements regarding that country to the greatest respect, has kindly supplied the Committee (October, 1882) with notes in reply to inquiries addressed to him for information, which we have condensed, thus:

ON THE SORGHUM (KOW-LIANG) OF CHINA.

1. About the year 1857 the French consul at Shanghai, M. de Montigney, introduced the Barbadoes millet (*Kowliang* of the Chinese) into France. He obtained the seed from the island of Tsung-ming, which lies in the mouth of the Yangtsz' Kiang, formed from deposits of detritus. This plant was exceedingly rich in juices, and when subjected in France to processes of manufacture produced a great quantity of saccharine matter. The plant attracted the attention of agriculturists in America, and they obtained small supplies of its seed from France, about the years 1855-'57, for purposes of making sugar. No seed was ever, to my knowledge, brought directly from China to this country.

2. The extraordinary richness of this plant grown on the island of Tsung-ming—resulting doubtless from the peculiarly fertile soil of this spot—is by no means equalled in other parts of China. All Chinese sugar is made from the sugar-cane (*Saccharum officinarum*) grown in the southern provinces, where sorghum is not found. From the latter the Chinese have never extracted sugar. In 1865 the United States Government sent an agent, Varnum D. Collins, to ascertain the methods employed among the Chinese in extracting and granulating sugar. This gentleman experimented upon the Tsung-ming sorghum seed, and obtained therefrom a sugar juice which considerably surprised the natives, who were wholly unacquainted with its saccharine properties.

The Chinese are abundantly supplied with good and cheap sugar in all portions of their empire, coming from the sugar-canes of the south; they have, consequently, no need of other sources than this plant. Their uses for sorghum are various: fodder for cattle, from its leaves; fuel, wattles for fences, &c., from the stalks. In binding several of these together and cementing with clay, they get a cheap substitute for posts, while the stalks in many ways take the place of timber.

Many varieties of the grain, black, red, and white, are known to the farmer. Its seeds, which are abundant, are used for making a sort of spirits, also occasionally for feeding to horses, mules, and camels.

This plant is almost wholly confined in its cultivation to the provinces north of the Yangtsz' River, and forms in this region one of the principal crops. It is not employed as food for man, save in times of famine and great stress. When ripe, the grain is about the size of duck-shot.

Question. Is it known how long sorghum has been cultivated in China as food, or for making spirits?

3. To this question it is hard to make any satisfactory reply, inasmuch as no Chinese books contain illustrations of grains or plants used in ancient times, nor are there found among their monuments pictures of these similar to representations of ancient Egypt, Assyria, Greece, &c.

As to the history of this grain in China, Dr. Bretschneider, of the Russian legation at Peking, and foremost among the authorities upon Chinese botany, says (concerning the plant called *Shu*): "This cereal is separately described in the *Pun Tsao* (Chinese Herbal), published A.D. 1570. The grain is called *Hwang-mi*, and is said to possess much glutinous matter. It is used for manufacturing alcoholic drinks. This corn was known to the Chinese in the most ancient times. It seems to me that the meaning of the character *Shu* in ancient days was not glutinous millet (as Dr. Legge states in the *Shu King*), but rather *sorghum*, as Dr. Williams translates."[*] If this deduction is true, the cultivation of this plant dates from about 2000 B. C. The precise uses of this grain in ancient times can only be inferred.

If the identity of the *Shu* (mentioned in the classics) with sorghum could be proved beyond question, this grain would rank in age as grown in China with any in the world.

4. Sorghum is seldom used in China now as food for man; the great food staples of Northern China are wheat, pulse, maize, and Italian millet (*Setaria*). Buckwheat, panicled millet, and the sweet potato may be included as secondary staples. Rice is imported to the north from the southern provinces.

5. I have never seen the broom corn grown in China.

6. The twenty or more varieties which President Angell brought from China could, probably, be increased in number if the collection were made from a more extended area.

The uses of this plant for fuel tend to increase attention to the development of its stalk rather than the grain.

The plant often attains a height of 15 or 16 feet. The common practice of stripping off all the leaves within reach upon the growing stalk, for feeding cattle, increases very materially its woody fiber. Cutting the stems while in their prime of growth, and chewing them green, as Southerners do the sugar-cane, is not unusual in the north.

The Chinese do not possess the art of refining sugar or making sirup to perfection. Even in cane-growing districts their employment of molasses is small; none of this is ever made from sorghum, to my knowledge.

Dr. E. BRETSCHNEIDER, physician to the Russian legation at Peking, who is quoted in the foregoing notes from Dr. Williams, says in his essay, or memoir, on the study and value of Chinese botany,[†] page 46 :

The true sugar-cane (*Saccharum officinarum*) growing in China must not be confounded with what is called *Northern China Sugar cane*. This is *Sorghum saccharatum*, a plant now a days largely cultivated in Europe and America for the purpose of manufacturing sugar from it. This plant was first introduced from Shanghai into France by the French consul, M. Montigney, in the year 1851, whence it spread over Europe and America, after it was proved that it is very rich in sugar.

Dr. Bretschneider then relates substantially the same statements, respecting Mr. Collins astonishing the natives by making sugar from sorghum, which Dr. Williams has already given.

On page 45, after discussing the meaning of the Chinese terms applied to these plants, he adds, in conclusion:

It seems to me that the meaning of the character translated *Shu* in ancient times was not glutinous millet (as Dr. Legge states in his translation of the Shu King), but rather sorgho, as Dr. Williams translates.

It seems, then, that the term *Chinese sugar-cane* is a misnomer only so far as the plant was not recognized as a sugar-producing plant by the Chinese, while the original seed of the *Sorghum saccharatum*, according to these authorities, was undoubtedly imported into France from China.

---

[*]As to the sugar-cane, the same writer adds: "I have not been able to find any allusion to it in the most ancient of Chinese works (the five classics); it is first mentioned by writers of the second century B. C. * * * One says, 'it grows in Cochin China; it is several inches in circumference, several *chang* (10 feet) high, and resembles the bamboo. The juice expressed is very sweet, and, dried in the sun, changes into sugar.'"

Sugar-cane is not mentioned as indigenous to China. The *Pun Tsao* (xxxiii, 13) gives a good description of the sugar-cane and its varieties, of the manufacture of sugar, &c., and quotes several authors of the Liang, Tang, and Sung dynasties, who describe the plant. In another book we learn that the Emperor, in A. D. 640, sent a man to India to learn there the method of manufacturing sugar.

[†]On the study and value of Chinese botanical works, with notes on the history of plants and geographical botany, from Chinese sources, illustrated with eight Chinese wood-cuts, dated Peking, December 17, 1870. 8vo. pp. 51. Printed by Rozario, Marçal & Co., Foochow.

2.—*M. LOUIS VILMORIN ON SORGHUM.*

M. Louis Vilmorin, of Paris, the well-known seedsman, in 1854, published in the *Bon Jardinier* Almanac for 1855, pages 41–53, an article on *Sorgho sucré* of much interest, from which it appears that sorghum was grown as a sugar plant at Florence, in 1766, by Pietro Arduino, and also that M. d'Abadie sent to the Museum in Paris from Abyssinia a collection of seeds containing thirty varieties of sorghum, some plants of which attracted attention from the sugary flavor of their stems. M. Vilmorin calls attention especially to the fact that while the seeds of sorgho from the new importation of Montigney from China in 1854 (see Dr. Williams's notes on the Chinese sorghum, above) were black and apparently identical with those of the old collections, the seeds of the Florentine plants were described as of a clear-brown color, corresponding to well-recognized differences in the sugar sorghum.

M. Vilmorin's article contains so much of interest as bearing upon the early history of sorghum, as well as results of well-conducted experiments by him to determine its industrial value for various purposes, that we add a translation of the paper to this Appendix. It is interesting to see how closely most of M. Vilmorin's results compare with those of Dr. Collier.

[Referred to in the report, p. 59. Translated from *"Le Bon Jardinier"* Almanac, Paris, for 1855, pp. 41–52.]

Sorgho sucré, Holcus saccharatus, Hort.; *Andropogon saccharatus?* Kunth.

This graminaceous plant, which seems destined to take an important place in the list of our industrial plants, was, like the "Igname of China," imported by M. de Montigny among other articles addressed in one sending to the Geographical Society.* We still hesitate about the botanical name by which this plant should be designated. The name of "*Holcus saccharatus*" is evidently erroneous, for, although the plant is very probably the same which was formerly so denominated, that division of this genus, which is characterized by the presence of a small pedunculate male awn by the side of each fertile awn, has been thrown out of the "*Holcus*" group into the genus "*Andropogon*" or "*Sorghum.*" In all likelihood the species "*Sorghum vulgare*" (*Andropogan sorghum*) will include among its varieties the plant which is now engaging our attention, as well as the "*A. cafer bicolor,*" &c., of Kunth. A recent work, yet unedited, which Mr. Wray, the author, has been kind enough to show to me, points out some fifteen varieties of this plant growing on the southeast coast of Caffraria, and in a collection of seeds from Abyssinia sent to the Museum, in the year 1840, by M. d'Abadie, and containing about thirty different kinds of sorgho, we had ourselves noticed some plants particularly remarkable on account of the sugary taste of their stalks. It is evident from all this that the occasions for confusion, which furnish at the same time a subject for critical examination, are not at all wanting. My colleague, Mr. Groenland, has, at my request, set about making a special study of the subject, and I hope that his researches, aided by the comparative cultivation of the several known varieties, will enable us to bring these different varieties back to the botanical types from which they were derived. Meanwhile we may just as well adopt the name "*Holcus saccharatus,*" which, although doubtless inexact in regard to the generic characters, has the advantage of being known and of never having been applied to other plants.

The plant which was submitted to the experiments made at Florence for the purpose of making sugar in the year 1766, by Pietro Arduino, belonged, very likely, to the same species, but it must have been of another variety, for he describes its seeds as light brown in color, whereas the seeds of the newly imported plant are black, and in all appearance identical with the "black sorgho" ("*sorgho noir*") of the old collections.

The "*sorgho sucré*" is a slender, tall plant, rising ordinarily to a height of 2 to 3 meters, and more on rich soils; its stalks are straight and glossy, the leaves flexuous and curved downward, and its general appearance is similar to that of maize, but

*See the *Revue Horticole,* February, 1854, "*Holcus saccharatus;*" July, 1854, "*Igname de Chine;*" "*Bulletin du comice agricole de Toulon,*" 1853; list of the Montigny sendings.

more graceful. As a rule, the sorgho forms a tuft of 8 to 10 stems, terminating in a conic panicle, thickly studded with flowers, green at first, and then changing through different shades of violet to a deep purple hue when they mature.

The plant is probably annual,* and its culture and time of growth agree with those of maize. In the climate of Paris it requires to be sown as soon as the soil is warm, viz, at the same time with the first seed-beds of kidney-beans. The maturity of the seeds is better assured when the plant has been grown in a sheltered nursery, or, still better, on a *deep hot-bed;* but, for the extraction of sugar, cultivation in the open field is sufficient, provided the soil be light and somewhat warm.

The product of the *"sorgho sucré"* consists of the juice, which is abundantly contained in the pith of the stalks, and which can furnish three important products, viz, sugar, alcohol, and a fermented beverage similar to cider.† In fact, this juice, if obtained with care, on a small scale, and stripping the cane of its green bark, is nearly colorless, and contains nothing, so to speak, but water and sugar. Its density varies from 1.050 to 1.075, and the proportion of sugar from 10 to 16 per cent. I mean here the total amount of crystallizable and uncrystallizable sugars, the latter amounting, sometimes, to one-third.

     *   *   *   *   *   *

Considered from the standpoint of the sugar manufacture, the sorgho has, as it seems to me, little chance of success in those regions, viz, the northern and central of France, where the success of the beet culture is already assured. The large proportion of uncrystallizable sugar which is contained in sorgho is not only lost for this industry, but it also creates a difficulty in regard to the extraction of the crystallizable sugar. What we mean is not, however, that the products of sorghum are poor, or difficult to be obtained, but simply that their nature renders them, all circumstances being equal, more important for alcohol than for sugar. If the distillation of beet, which does not yield, even by the most perfect processes, an amount of alcohol proportional to the quantity of sugar that can be extracted, gives, in the present condition of the market, a good profit, the sorgho juice, yielding much more in alcohol than in sugar, will, *a fortiori*, prove also profitable.

The result would, of course, be different for sorgho cultivated in warmer regions, and where beets cannot grow by its side. Some experiments made on some sorgho stalks cultivated in Algiers, and which had been sent to me by M. Peschard d'Ambly, the mayor of Philippeville, gave me a product in sugar considerably superior to that obtained from my own plants, grown near Paris. Owing to the length of time required for the parcel to reach me, an alteration had set in, which rendered it impossible for me to determine with certainty the ratio of the two kinds of sugar in the sorgho from Algiers. But the nature of the juice, and the observations communicated to me by Mr. Wray, an old colonist of Natal (Caffreria), lead me to think that the proportion of crystallizable sugar becomes much greater whenever the climate allows the sorgho to reach complete maturity. This plant, then, might fill, in regard to the production of sugar, the gap intervening between the tropical regions—the only ones adapted to the culture of cane—and the forty-fourth parallel, which seems to be the southern boundary of the belt where beet culture is profitable. The beet will, very likely, maintain itself in the field of sugar production beyond this limit, whereas the sorgho will surely prevail, chiefly in the western and southwestern provinces of France, as an alcohol-producing plant.

As a sugar-producing plant sorgo would have in its favor the facility of cultivation and of the treatment of its juices. Its gross product will probably surpass that of the sugar-cane in those countries where, as in Louisiana, *e. g.*, the latter becomes an annual plant. Its tops and leaves would also furnish abundance of excellent green forage. Finally, its molasses, wholly similar to that of cane, could be used for the production of rum, and its juice for that of a liquor very much like tafia. The main difficulty would probably be to preserve the stalks long enough to allow time for manufacturing. But, without saying that the climate within the above-mentioned geographical limits would permit of successive crops in the same season, I learn from Mr. Wray, whom I have already cited, that in the vicinity of Natal the Zulu Caffres‡

---

* I say probably, because when I saw, last autumn, the vigor and size of the stubbles I thought that they might be put in under shelter to furnish plants for the following spring.

† See the *"Moniteur universel"* of November 13, 1854; also the *"Revue horticole"* of November 16.

‡ The Zulu Caffres cultivate a great number of varieties of the sugar sorghum, (cal'ed by them "imphee"), not for the purpose of making sugar from them, but of sucking their stalks. M. Boussignault has told me of late that in New Grenada pieces of sugar-cane and f maize stalks are sold in the market-places for the same use. There is in this a suggestion quite interesting in regard to the *"sugar-maize question,"* which is now engaging me (see the *"Revue horticole,"* November 10, 1854, p. 426), and about which I propose soon to speak.

preserve the sorghum stalks for a very long time by burying them in the earth, which in that climate is very warm and damp.

We have just seen that sorgho has for the production of alcohol the advantage of turning to account its uncrystallizable sugar, which is lost when sorghum is exclusively used for the manufacture of sugar. Another advantage consists in the pureness of its juice, owing to which the alcohols, and even the crude brandies, obtained from it are pure enough to permit immediate delivery to consumers.

The alcohol, imperfectly rectified through one distillation only, and which I obtained with a laboratory apparatus altogether incomplete, had no foreign taste whatever; and even my products of from 40 to 50 per cent. had a rather agreeable taste, somewhat like that of "*eau de noyau*," and I have no doubt they could be used to mix with the analogous products of beet. When they are genuine their flavor is by far less strong and less peculiar than that of rum, and I am convinced that if permitted to age they would be excellent.

I have said above that the sorgho juice could furnish, besides alcohol and sugar, a beverage similar to cider. In a note in the *Moniteur* of November 13, 1854, as also in the *Revue horticole* of the 16th same month, I endeavored to call attention to the advantage there might be found in trying to make such fermented beverages. This question is yet entirely new; therefore no one can foresee whither it may carry us; but from the results which I obtained, under circumstances entirely unfavorable, I cannot help thinking it has some prospects.

The culture of sorghum is not likely to meet with difficulties; it will be conducted pretty much in the same way as that of maize and millets, and it, moreover, already exists in some of our provinces. The question of the place to be assigned to the sorghum, in the laying out of fields into plots for rotation, is probably the only one which is likely to give occasion to difficulties. In this respect I believe that there is more to lose than to gain by the introduction of the new crop, while beets have, on the whole, constantly improved the production of the regions where they have been cultivated. Sorghum is, as well as maize, considered as an exhausting plant. My experiments in this direction have, so far, been not numerous enough to enable me to form a correct opinion on this point on the basis of direct observation, but I have very good reason to believe that sorghum is really an exhausting plant; at any rate, the family to which it belongs makes it little probable that it should rank with beets in regard to rotation.

The estimate of the products that can be expected from sorghum is difficult enough in the present state of the question; my experiments so far have been on a very small scale. The following are the data gathered from them, and the conjectures that can be formed:

The small crop of sorghum which I had cultivated in an open field at Verrières, on a sandy soil of middling quality, was harvested on October 30. The cultivation had been proportioned to the wants of the different experiments in the laboratory; moreover. one portion of the crop was much impoverished by the vicinity of a large tree, and another was preserved for a comparative experiment on lopped plants mingled with others having their panicles on. The surface cut over on October 30 measured 58.40 meters. The produce was (weight taken on the day after the cutting):

	Kilograms.
Stalks and leaves	285.400
Stalks stripped of leaves and tops	179.250

The plot on which the crop had been raised was so irregular and uneven that I determined on taking a counter-proof. In that portion which lay at the greatest distance from the trees I traced a square wherein the plants were, if not fully developed at least all even, and the ground equally filled throughout.

The surface was 5^m.32, and its produce, weighed the day subsequent to the cutting, was as follows:

	Kilograms.
Stalks and leaves	41.110
Stalks stripped of leaves and tops	26.230

In my judgment, this portion of the crop could well be taken as representing a good average harvest, such as would be 45,000 kilograms of beets per hectare. It is on these figures that I am about to make my comparisons.

In the same portion of the field the plants had been forwarded on a hot-bed and transplanted in rows at the beginning of May.

In another part, where the ground had been sowed on May 18, only a few plants appeared. The crop had not been weeded when young, and its growth was, therefore, much retarded. Its average gross produce was 38,000 kilograms to the hectare.

A third crop, started on a hot-bed and transplanted in a garden, was not weighed; we took now and then from it what we wanted for our own experiments, directed to the determination of the period when sugar is developed. Yet I do not think I am mistaken in estimating this produce, from the appearance of the plants, as surpassing by a half the yield of the plot of $5^m.32$ given above.

The proportion of the juice obtained from stalks stripped of leaves was from 55 to 60 per cent. It is plain that by working canes which have been carefully selected' or severely topped, the yield must needs be considerably increased. With a good mill it should easily reach 70 per cent. The juice gathered from the treatment (carried on in the cider-press of the village on October 29) of 215 kilograms of large and small stalks, from which the ears and the last joint had been cut off, was 106 liters, the densimeter marking 1.052. I estimate the loss incurred in wetting up the extended surfaces of the trough and the press at 5 liters.

I have not extracted any sugar from sorghum; I have only made some determinations by means of the saccharometer, and verified them generally by means of evaporation and a treatment with alcohol.

The following are the results presenting the proportion of sugar existing in the juice from plants gathered at Verrières:

Per centum.

October 13, 1853	10.04
November 28, 1853	13.08
November 28, 1853, second experiment	14.06
October 13, 1854 (without inversion)	10.14
November 15, 1854, crystallizable sugar, 11¾ per cent. ; uncrystallizable sugar, 4¼ per cent	16.00

The amount of alcohol produced by the juice was ascertained by the direct method of fermentation. The following are the figures in the order in which they were determined (these figures represent the cubic centimeters of absolute alcohol per liter): *

Sorgho from Verrières: Cubic centimeters.

September 28, 1854	41.00
October 4, 1854	54.00

Sorgho from Algiers:

First fermentation, October 17, 1854 (with the Salleron apparatus)	70.00
First fermentation, second trial (with the Salleron apparatus)	74.00
First fermentation, second trial (by distillation of 1.20 liters)	70.72
Second fermentation, October 18, 1854	79.52

Sorgho from Verrières:

October 20, 1854 (by distillation of one liter)	72.51
November 16, 1854 (panicles cut)	63.26
November 17, 1854 (panicles preserved)	60.67

If we suppress the figures belonging to September 28, which refer to plants evidently too young, as also the four figures representing the sorgho from Algiers, we find that 6.3 per cent. in volume, or 63 cubic centimeters of alcohol per liter of juice, is, in our climate, the average figure, which plainly seems quite encouraging, especially considering the excellent quality of the product.

Our calculations, on the basis given above, would show that the returns of one hectare of sorghum would be as follows:

Stalks and leaves	kilograms	77,270
Net stalks	do	49,300
Juice, at 55 per cent. to the weight of stalks (271 hectoliters)	liters	27,115
Sugar, at 8 per cent. to the juice	kilograms	2,169
Absolute alcohol, at 63 per cent. to the juice	liters	1,708

The analogous returns from beets would be as follows:

Roots, weight to the hectare	kilograms	45,000
Juice, at 80 per cent. to the weight of roots	do	36,000
Sugar, at 6 per cent. to the juice	do	2,160
Absolute alcohol, at 3 per cent. to the beets	liters	1,350

The 8 per cent. sugar on which I have calculated the yield of sorgho will perhaps be considered as too low, but it should not be forgotten that it refers to the crystallizable sugar that can actually be extracted, and I do not, therefore, believe my estimate too low. If I were to make a comparison between the "sorgho" and the "sugarcane" in a more southerly climate, I have no doubt that the figure representing the product in sugar would rise to a far higher value; but I lack the data required for such a comparison, as well as for a comparison between the same plant and the vine, or the Jerusalem artichoke ("topinambour"), or the grains, or even the daffodil ("asphodèle"), in respect to the production of alcohol.

After examining the chances of the industrial culture of sorgho, and the considerations that may lead to the adoption of this plant, I have only to furnish some data obtained from our first experiments, which may afford some indications for further study, or some guide for the first attempts in manufacture.

One of the points which I have endeavored to establish, without, however, obtaining complete success, was this, viz: What is the time, during the period of vegetation, when the stalks begin to contain sugar, and, consequently, what is the moment when the manufacture may commence? It appeared to me that this time coincided with that of the appearance of the ears; but the proportion of sugar existing in the cane keeps on increasing up to the time when the seeds pass into the milky stage. I have not ccd that the richness in sugar in a plant while blooming diminished gradually from the lower to the upper part of the stalk in the spaces between the joints, and also that the lower portion of each one of these interspaces is younger and less rich in sugar than the upper one. Such being the case, the middle of the stalk is the ricl est portion, for the lower joints are hard and small. I have not been able to ascertain it with exactness, but I suppose that at a later period the spaces botween the joints in the lower part of the stalk become impoverished, or, if the juice does not grow poorer, it at least diminishes in quantity.

The ripeness of seeds does not seem to reduce to any considerable degree the production of sugar, at least in our climate; but as maturity is reached at the end of the season, and our plants, consequently, continue to advance in richness with the development of vegetation, the effect of maturity on these phenomena can hardly be determined. This question can be solved only in those countries where the seeds of the plant mature before the warm season is over. According to M. de Beauregard's report, addressed to the "Comice de Toulon," maturity would seem to have had no injurious influence within the limits of his experience; and he considers seed and sugar as two products which can be obtained jointly. On the other hand the Zulu Caffres are accustomed to snatch, by an abrupt pull, the panicles away from their plants as soon as they show themselves, in order to increase the sugary quality of the stalks. But this question has, after all, no importance in respect to France, since here ripeness will never take place too soon to prove detrimental.

3.—LETTER FROM MR. LEONARD WRAY.

Mr. Leonard Wray is the veteran pioneer of sorghum culture in the United States and in France, whose contributions have already been referred to in the body of this report. His early communication in the agricultural section of the Patent Office Report, Part III, 1854, p. 219, will be read with much interest in this connection. His observations are there reported by D. Jay Browne. He now writes as follows:

PERAK, via PENANG, September 7, 1882.

To the Commissioner of Agriculture, Washington, U. S.:

DEAR SIR: I am pleased beyond measure to find that the United States Government has *at last* awakened to the great value of the "imphee varieties," which I introduced into your country, and has taken the most certain course to verify by scientific tests the *truth* of my printed statements respecting them, published in English, and also in French, in 1854, copies of which I gave to Mr. D. J. Browne, of the Patent Office, in Washington.

You will find the contents of this, my pamphlet, in a little book by H. S. Olcott, published by Moore, of Fulton street, New York, in 1857; and if you do me the honor to read that, you will, I am sure, fairly acknowledge that every statement I therein made is *strictly proven* by the valuable results of the able men whom you selected to conduct your experiments. I must, however, mention that the *last chapter* of my pamphlet, viz, that on the manufacture of the imphee juice into sugar, is omitted in Olcott's little book.

It is most gratifying to see the "*thorough*" manner in which your Department has gone into and decided these important questions.

I first became acquainted with these plants in March, 1851 (thirty and one-half years ago), just after my arrival in Natal, South Africa; and in 1854 I grew them in several parts of France, in England, Spain, Italy, and in various other places, so that I may claim to know their merits, and I *now* say that all I said and wrote about them at that time I am fully prepared to stand by and substantiate the truth of.

In fact, your admirable Department has, in its recent scientific demonstrations, abundantly and authoritatively confirmed my facts, and thereby rendered an inestimable service to your country, and to other countries also. I hope and trust you will continue it.

Looking at the beautiful plates in your reports, I cannot but express my admiration, and at the same time my astonishment, at the very remarkable *constancy* of the "types" maintained by the different sorts of imphee shown. For instance, I may mention Plate I, facing page 8, in S. Report 33. This is there called "Imphee Liberian" and "Sumac"; but I distinctly recognize it as my "Koom-ba-na," one of the very sweetest and best I had. (I inclose you some *very old* seed.)

Plates 2, 3, and 4 are my Neeñzãnã and its sports.

Plate 5 is my En-yã-mã, which I see figures as "W. Mammoth." I inclose some of my old seeds of it.

Plate 7 is my Oom-see-ã-nã.

Plate 8 seems to me to be the "Chinese sorgho."

Plate 9 is an Oom-see-ã-nã kind (no doubt a "sport").

Plate 10 is undoubtedly my "Vim-bis-chu-a-pa," which, to please General Hammond, I nicknamed Sorgho Ka-bai (or Sorgho Brother). Some grew to 6 pounds weight when "topped," and I had the head of one such until about nine months ago, when I unluckily threw it away (it was 20 inches long). I see you call it by the names of Honduras, Honey, Mastodon, &c.

Plate 11 seems to me to be no other than my Boom-vira-na, one of my special favorites. Please see the description in my little pamphlet (in Olcott's book, 1857), and I think you will not long be in any doubt about its origin, bogus stories notwithstanding.

Plates 12 and 13 are both my imphees, and I had some growing here twelve months ago, but the seed unfortunately got spoiled.

The seed you were kind enough to favor me with I have sown and had sown by my friends here; and mine are now 8 inches high, being only sixteen days' growth. I may mention that I soaked my seed in a strong solution of sugar with a little *salt, camphor, and soap-suds* for twenty hours, and I think they are growing much more vigorously than those *not* so treated. I shall continue to watch them.

Pray do not think me ungrateful when I say that I felt disappointed in not finding any "*Minnesota E. Amber,*" nor any "Oomseeana" amongst the seed you sent me, and I trust you will forgive me if I trespass so far on your kindness as to beg that you will be so good as to send me some of those two kinds, also "*White Mammoth*" and "*Sumac,*" all of which I particularly wish to have. Even 100 or 200 seeds of *each of these four sorts* will be ample for me to propagate from, and these might come in a letter direct to me here (and not by Singapore). In such case the correct address is: "Perak, via Pinang, straits of Malacca," nothing more.

I need not say, also, how thankful I shall feel for any of your instructive reports or other information you may be kind enough to bestow upon me.

I will by no means neglect to send you a goodly assortment of such seeds as I think you will be glad to have, as soon as they are ready. With many excuses for so troubling you, I beg to subscribe myself, dear sir,

Yours, very faithfully,

LEONARD WRAY.

4.—*FACTS REGARDING SORGHUM, AND SOME CONCLUSIONS AS TO ITS VALUE AS A SOURCE OF SUGAR.*

By PETER COLLIER, PH. D., *Chemist: United States Department of Agriculture.*[*]

Having given considerable study during the past four years to the sorghum plant, I take this occasion to present to the scientific public a brief and necessarily somewhat incomplete *résumé* of work accomplished, together with such conclusions as seem warranted by the facts.

Botanical definition.—The genus *Sorghum* (of which *Sorghum vulgare* is the accepted type) is included in the natural order *Graminaceæ*, to which natural order belongs also the tropical sugar cane (*Saccharum officinarum*); but it should be remarked that between the genus *Sorghum* and the genus *Saccharum* there are classed by botanists the three genera, *Erianthus, Eriochrysis*, and *Ischæmopogon*. Vid. Grisebach's "Flora of the West India Islands," pp. 560, 561.

While, therefore, the two plants are somewhat closely related, this relationship does not warrant the assertion made by a recent writer upon this subject, "that the name sorghum is a mere disguise, for the reason that it is nothing more nor less than

---

[*] This paper was transmitted to the Sorghum Sugar Committee March 26, 1882, by President Rogers, agreeably to a request from the Hon. Commissioner of Agriculture, of date March 24. It is the communication submitted by Dr. Collier to the Academy, on invitation, at the session held in Philadelphia November, 1882.

a subvariety of sugar-cane, which may explain why it is that the reader and the investigator have so frequently been misled."

To the unscientific observer a growing sorghum plant would seem to combine many of the exterior characteristics of sugar-cane.

Subvarieties.—There seem to have been originally introduced into this country two principal types, or subvarieties of sorghum, viz, the African and the Chinese. At present it is difficult to give more than an approximate estimate as to the number of subvarieties actually cultivated. During the season of 1880 there were grown on the land of the Department of Agriculture, at Washington, thirty-eight subvarieties (several of which, though of different names, seem to differ botanically very little, if at all), and within the past month I have received directly from Natal the seeds of thirteen subvarieties there grown, which, so far as I am able at present to judge, do not correspond with any which I have previously examined.

From the fact that these subvarieties hybridize quite readily I am led to infer that the number of distinguishable types cannot be far from seventy.

As would naturally be expected, these different subvarieties vary considerably as regards external appearance, size, height, length of season required for complete development and maturing of the seeds, and consequently for the development of the maximum amount of crystallizable sugar.

POINTS OF AGREEMENT FOR SUBVARIETIES.

With but one or two anomalous exceptions all the thirty-eight varieties which I last year examined (and this years' work is confirmatory) agreed in the following points, viz:

1. *Soil required.*—All varieties do well on soil of average fertility; bottom lands raise the finest canes.

2. *Heat and moisture effects.*—Like corn, considerable rain is advantageous after the plant has well begun its growth, provided, also, the following weather be quite warm.

3. *Time of reaching maximum sugar content.*—All varieties reach a maximum sugar content at, or about, the time when the seed is fully matured.

4. *Maximum in different varieties.*—This maximum content is practically the same for the different subvarieties; it may be said in average years not to vary greatly from 16 or 17 per cent. of crystallizable (cane) sugar in the juice. During this season, which has been exceptionally dry, the percentage of juice extracted has been somewhat smaller, and maximum increased by dry season; in consequence, the percentage of sugar in the juice has been increased to a maximum of 18 or 19 per cent.

5. *"Exponent."*—The average purity ("exponent") of the different juices at maturity of the canes seldom falls below 70 per cent., and frequently exceeds 80 per cent. This "exponent," as I have termed it, is obtained by dividing the total cane-sugar in the juice by the weight of the total solids, the latter being determined by drying a given weight of the juice with sand, at 90° to 100° C.

6. *"Available sugar."*—The "average available sugar" in the juices of matured canes varies from 8 to 13 or 14 per cent. of the weight of the juice, the lower figure being an average for all varieties while the higher figures are for the best varieties. This available sugar may be obtained by subtracting from the total solids the sum of the glucose and the "solids not sugar." Or, with juices from mature canes, practically the same results are obtained by multiplying the percentage of cane-sugar in the juice by the "exponent." The first method of determining "available sucrose" is applicable to all juices; the second gives practical results where the "exponent" exceeds 65 or 70 per cent.

7. *Life history.*—The life history of the different subvarieties of sorghum, so far as the composition of the juices can throw light upon the subject, is very similar for all, except that the actual time for reaching maturity varies considerably.

The following table, deduced from the results of 2,739 analyses of sorghum canes, presents, in a condensed form, a very correct idea as to the actual development of the cane itself and of the changes in the juice:

Table showing general averages for the stages, as determined from the results of the same stage for all varieties of sorghum.

Stages.*	Average length.	Diameter.	Unstripped weight.	Stripped weight.	Per cent. of juice.	Specific gravity.	Per cent. glucose.	Per cent. sucrose.	Per cent. solids.	Exponent.	Per cent. available sucrose.	Number of juices analysed.
1	7.5	0.9	1.93	1.34	59.96	1.033	4.39	1.76	1.75	22.56	0.40	58
2	8.5	.9	1.93	1.46	59.00	1.038	4.45	2.96	1.80	31.93	.95	69
3	8.8	.9	1.73	1.39	59.67	1.037	4.50	3.51	1.78	35.85	1.26	57
4	9.1	.8	1.83	1.44	61.61	1.041	4.34	4.34	1.91	44.98	1.78	70
5	9.3	.9	1.96	1.55	63.05	1.045	4.15	5.13	1.92	45.80	2.35	75
6	8.7	.9	2.02	1.60	62.79	1.050	3.99	6.60	2.45	50.23	3.26	62
7	9.7	.9	2.11	1.55	63.85	1.052	3.86	7.38	2.19	54.95	4.06	70
8	9.3	1.0	2.10	1.63	65.68	1.055	3.53	7.60	2.37	55.36	4.26	111
9	8.8	.9	1.87	1.40	64.88	1.058	3.19	8.95	2.42	61.47	5.50	306
10	8.9	.9	1.81	1.38	64.83	1.061	2.60	9.98	2.50	65.18	6.68	217
11	9.1	.9	1.94	1.48	65.92	1.063	2.35	10.06	2.73	67.77	7.22	166
12	9.0	.9	1.81	1.37	63.39	1.065	2.07	11.18	2.63	69.53	7.77	170
13	9.1	.9	1.86	1.34	62.90	1.066	2.03	11.40	2.82	70.15	8.00	183
14	8.9	.9	1.82	1.32	61.72	1.067	1.88	11.76	2.90	70.84	8.53	191
15	8.9	.9	1.81	1.32	60.45	1.067	1.81	11.60	3.15	70.21	8.21	217
16	8.7	.9	1.73	1.22	61.30	1.070	1.64	12.40	3.33	71.43	8.86	329
17	7.7	.9	1.60	1.25	60.17	1.073	1.56	13.72	4.07	74.90	9.73	197
18	8.5	1.0	1.44	1.15	62.00	1.069	1.85	11.72	3.48	69.34	8.17	191
19†	8.5	.9	1.81	1.52	56.04	1.080	2.00	12.06	3.63	64.70	7.62	30

Among the points of most practical interest may be mentioned the following:

1st. The changes in height, weight, diameter, and total and stripped weight are not sufficiently important to require comment.

2d. The percentage of juice extracted from the stripped stalks gradually increases up to the eleventh stage, then slowly diminishes until the close of the season.

3d. The specific gravity of the juice, the percentage of sucrose, the percentage of solids not sugar, and the exponent regularly increase (with but one or two exceptions) until the close of the season; and the percentage of glucose in the juice as steadily decreases from the first.

It will here be noticed that the sucrose increases in the juice much more rapidly than do the solids not sugar; and this fact, taken together with the steady decrease of glucose, is the explanation of the equally steady increase of the exponent, which represents the comparative purity of the juices.

---

* The "stages" to which reference is here made are defined as follows:

SORGHUM.

Stage.	Development of plant.
E	About one week before opening of panicle.
F	Immediately before opening of panicle.
1	Panicle just appearing.
2	Panicle two-thirds out.
3	Panicle entirely out; no stem above upper leaf.
4	Panicle beginning to bloom on top.
5	Flowers all out; stamens beginning to drop.
6	Seed well set.
7	Seed entering the milk state.
8	Seed becoming doughy.
9	Seed doughy, becoming dry.
10	Seed almost dry, easily crushed.
11	Seed dry, easily split.
12	Seed split with difficulty.
13	Seed split with more difficulty.
14	Seed split with still more difficulty.
15	Seed harder.
16	Seed still harder.
17	Seed still harder.
18	Seed still harder.

† This stage (No. 19) was after the cane had ceased growing, late in the season; it was determined from canes Nos. 23 and 24 only.

Having given some points wherein the various subvarieties of sorghum resemble each other, I may state that they differ more or less distinctly in the following: Botanically they differ decidedly.

(a.) *Shape of seed head.*—In the form of the seed head, which is loose and spreading in the Honduras varieties (*i. e.*, "Honduras," "Mastodon," "Sprangle-top," "Honey Cane"), quite compact in the Liberian varieties (*i. e.*, "Liberian," "Imphee," "Sumach"), on a recurved stalk in "Rice" or "Egyptian corn," and with various intermed iate modifications for other subvarieties.

(b.) *Size and appearance of seeds and glumes.*—The proportionate sizes of seeds and glumes, as well as the size and color of the seeds, and the color and adherence or nonadherence of the glumes, are all strikingly noticeable.

(c.) *Height, weight, and diameter of stalks.*—The height, weight, and diameter of the stalk s are very different; thus the average above named are for the Honduras varieties considerably greater than for the varieties termed "Early Amber," "Liberian," "Sumach," "Imphee," &c. This statement is based upon a study of Tables Nos. 1 to 38 of my last report on analyses of sorghum. (Department of Agriculture, Special Report No. 33 (1881), pages 1–55.)

(d.) *Time from planting to maturity.*—The varying rapidity with which the different varit ties come to maturity is one of the most striking physiological peculiarities; thus three samples of Early Amber seed from different sections of the country, and plan ed at the same time, required, respectively, 77, 80, and 89 days for complete deve opment; "White Mammoth," 102 days; three samples of "Oomseeana" from different sections, 104, 115, and 127 days; Chinese, 137 days; three samples of Hondura s, 148, 157, 164, &c. (*Vide* page 96, Special Report, No. 33, Department of Agriculture, 1881.) Excepting one sample of Honduras, all the sorghums mentioned, together with many more, were planted at the same date, in the same field, and had as n arly equal conditions of soil and treatment as could be afforded.

These facts show the necessity for discrimination in the selection of varieties, in order that such may be grown as shall prove well adapted to the climatic conditions of the region where they are to be introduced. Other things being equal, the quick-maturing varieties are best adapted to those northern latitudes which have a short summer season. The secret of the success of the Early Amber cane in the North is, in my opinion, the fact that it matures quickly, thus attaining its maximum sweetness long before serious danger of frost, rather than to any peculiar property of "granulating well," or to any greater content of sugar. For the same period of development this variety cannot be pronounced better than many others.

On the other hand, reports from Texas and South Carolina convince me that the Honduras varieties may, with the long seasons there possible, be better, for the reason that there they do mature, then weigh nearly or quite twice as much per stalk as does the Early Amber, the juice is of equal purity and sugar content, and hence, with soil of equal fertility, nearly twice the average northern crop may be secured.

A very valuable variety known as Link's Hybrid, originally from Tennessee, requires a little too long a season perhaps (101 days observed) to allow it to compete in the more northern States with the varieties which mature more quickly, but it is remarkably well adapted to the conditions of soil and climate in the section from which it came.

(e.) *Working period.*—Another very important consideration is the length of time after a cane first reaches maturity that the season and the habits of the plant conspire to preserve the juice in condition fit for working; in other words, the longer the "working period" the better.

The average length of time from planting to the death of the cane from frost seems to vary, in Washington, from 180 to about 200 days—six or seven months. As a rule, it appears that those canes which mature first ("Early Amber," "Early Golden," "Golden Syrup," &c.) continue to furnish juices of good, workable quality up to the time of their death from frost. Hence in this latitude these are the ones to be preferred.

I have stated very briefly some of the principal points of practical importance relating to the physiological development of this interesting plant. Many of these points, though seemingly simple, were ascertained only through long and patient investigation. The averages here presented are based upon an aggregate of from 2,500 to 4,000 analyses of the growing plants, and they are further strengthened and confirmed by the results of over 1,800 analyses executed this year. In other words, it is hard to conceive of any great departure from truth in facts based upon such a considerable number of determinations. All smaller personal errors or errors inherent in analytical processes are likely to be partially or entirely eliminated; at all events, these errors cannot exceed, on a juice containing 15 per cent. of cane sugar, $+0.2$ or -0.1 per cent., as has been demonstrated by a considerable number of careful experiments. Very few accepted facts of physical science rest upon a greater number of determinations.

It appears, then, that for the whole working period of those varieties more or less well adapted for cultivation in Washington, the average composition of the juice was about as follows:

	Per cent.
Cane sugar	13.0
Glucose	1.5
Other organic solids	2.0
Ash	1.0
	17.5

The percentage of purity ("exponent") being not far from 74 per cent., and the cane sugar in excess of all impurities being about 8.5 per cent.

These being averages for a long period, and including all the canes that may be said to give any promise of usefulness in that latitude, cannot be considered as the maximum results attainable with the best varieties. In fact, quite a number of varieties furnish juices which contain, during a very considerable number of weeks, fully 3 per cent. more available sugar than above stated; the composition of these juices being about as follows:

	Per cent.
Cane sugar	16.0 to 15.5
Glucose	0.6 to 1.0
Other organic solids	2.0 to 2.0
Ash	1.0 to 1.0

Or an amount of "available sugar" ranging from 11.5 to 12.4 per cent. That juices of this character will prove valuable for the production of sugar and sirup I cannot doubt. That many practical failures have resulted from the inexperience and lack of knowledge of over-sanguine experimenters is not a matter for surprise; the greater wonder is that any decided success should have been had within such a short time by novices in this branch of industry.

I am happy to be able to present for your inspection two letters, one from Mr. Porter, of Red Wing, Minn., the other from Mr. A. J. Russell, of Jaynesville, Wis. (See end of this exhibit.)

During the season of 1880, Mr. Porter made 4,000 pounds of marketable sugar which he sold at 9 cents per pound; during the same season Mr. Russell produced over 14,000 pounds of sugar from one-third of his crop of molasses (the remainder of which was equally good); this sugar sold in Chicago for 10, 9¼, and 9 cents per pound according to quality, and the molasses brought in the same market 50 cents per gallon in car lots, and 55 cents per gallon in five-barrel lots.

Both these gentlemen were embarrassed by inadequate machinery, but their decided success has encouraged them to continue as soon as better apparatus may be obtained.

I claim that in the infancy of an industry requiring so much knowledge of manufacturing methods, a few instances of conspicuous success should have more weight in the minds of scientific judges than a considerably greater number of failures.

It should be remembered that time is required for the diffusion of practical knowledge, and that beginners are frequently but poorly prepared for the practical difficulties they are sure to encounter. Were a novice to be placed in charge of Bessemer-steel works, success could hardly be expected; in like manner the management of a process of manufacture little less difficult, and which deals with organic substances, can hardly be left with safety to inexperienced men.

Thus far I have presented facts.

In conclusion, I would say that the judgment of a practical sugar maker of fifteen years' experience (Mr. Peter Lynch, of Baltimore, Md.), is to the effect that with selection of proper varieties, good soil, good cultivation, and proper handling of the juices and sirups, the sum recoverable for sugar and sirup will be such as to yield a profit greater than could be expected from corn, growing during the same season upon the same soil.

I present for your examination a considerable amount of printed and manuscript evidence bearing upon the points which I have already stated.

I shall be pleased to have a most careful investigation by a committee of practical chemists, members of this academy. I feel assured that the magnitude of the interests at stake warrants me in thus asking an impartial verdict from men of acknowledged fairness and ability.

NOVEMBER, 1881.

The letter of Mr. J. F. Porter, of Red Wing, Minn., will be found in the Appendix p. 123; the letter from Mr. A. J. Russell, of Janesville, Wis., is as follows, viz:

JANESVILLE, WIS., *November* 4, 1881.

DEAR SIR: Your favor of the 31st of October at hand.

Before the centrifugal broke, we had 14,752 pounds of dry commercial sugar, about like that we sent the Department and President Hayes. We had two-thirds of our melada left over until next season. By careful computation of what remained on hand' with the number of graining tanks emptied, and there was 44,256 pounds of dry commercial sugar. Most of it sold for 10 cents per pound. We sold some at 9 to 9½ cents.

In 1880 the juice was poor, but we made here a sirup that sold to the jobbers in Chicago at 50 cents per gallon by the cargo.

We have no machinery that we think adapted to sugar making profitably, and have confined ourselves to sirup, which we sell at 55 cents by the 5-barrel lots, &c. All my experiments on the stove for sugar this season were satisfactory in sugar. Will put in sugar machinery next year.

Respectfully, yours,

A. J. RUSSELL,
Formerly of the firm of Waidner & Russell, and manager of the C. L. Sugar Works.

To PETER COLLIER, Esq.,
Chemist, Agricultural Department, Washington, D. C.

5.—REPORT UPON STATISTICS OF SORGHUM.*

HON. GEO. B. LORING:

SIR: I respectfully present, in accordance with your requirement, sundry facts showing the status of sorghum production, from the census of the United States and from State enumerations.

It will be seen great fluctuations in area have occurred, that the greatest extent of cultivation in the older States was during the war period, and that a decline followed, except in newer States rapidly advancing in settlement.

There is scarcely any record of sugar, except in Ohio, where the product was greatest prior to 1870.

In the more western States there has been a revival of interest and extension of cultivation since the introduction of the Early Amber variety, from which some sugar has been made.

In 1860 and 1870 the census presented production as follows:

	1870.	1860.
	Gallons of sirup.	*Gallons of sirup.*
Indiana	2,026,212	881,049
Ohio	2,023,427	779,076
Illinois	1,960,473	806,589
Kentucky	1,740,453	356,705
Missouri	1,730,171	796,111
Tennessee	1,254,701	706,663
Iowa	1,218,636	1,211,512
Product of the above States	11,954,073	5,537,705
Product of other States	4,096,016	1,211,418
Product of the United States	16,050,089	6,749,123

The returns of sorghum in the recent census have not been tabulated except in two or three States. Only South Carolina and Kansas are complete, as follows:

	1880.			1870.
	Acres.	*Pounds sugar.*	*Gallons molasses.*	*Gallons molasses.*
South Carolina	7,660	8,225	276,046	183,585
Kansas	25,643	18,060	1,414,404	449,409

* This document was transmitted to the committee March 26, 1882, at the request of the honorable Commissioner of Agriculture. The returns for 1882, being incomplete, are not included.

The following States, in which the interest has been and is most prominent, are thus represented by local official enumerations:

OHIO.

Years.	Acres.	Sugar.	Sirup.
		Pounds.	*Gallons.*
1862	30, 872	27, 486	2, 690, 159
1863	31, 255	27, 359	2, 347, 578
1864	29, 392	41, 660	2, 609, 728
1865	37, 042	56, 666	4, 003, 754
1866	43, 101	46, 951	4, 629, 570
1867	17, 804	20, 094	1, 255, 807
1868	25, 257	28, 668	2, 004, 055
1869	22, 231	27, 048	1, 683, 042
1870	23, 450	21, 988	2, 187, 673
1871	23, 072	25, 505	1, 817, 042
1872	12, 932	34, 599	968, 139
1873	9, 426	36, 846	692, 314
1874	12, 106	36, 410	941, 510
1875	13, 144	21, 768	928, 106
1876	15, 929½	25, 074	1, 185, 235
1877	16, 104½	7, 507½	1, 180, 255
1878	16, 305	11, 909	1, 273, 048

MINNESOTA.

(No official returns of sugar.)

Years.	Acres.	Sirup.
		Gallons.
1868		81, 375
1869	629	31, 191
1870	728	56, 370
1871	1, 244	73, 425
1872	859	78, 005
1873	747	53, 226
1874	1, 146	69, 599
1875	1, 534	70, 479
1876	1, 695	72, 489
1877	2, 200	140, 153
1878	3, 207	329, 660
1879	5, 033	446, 940
1880	7, 317	....

IOWA.

Years.	Acres.	Sirup.	Sugar.
		Gallons.	*Pounds.*
1865	21, 452	1, 443, 605	8, 386
1867	25, 796	2, 094, 557	14, 697
1869	26, 243	2, 592, 393	....
1875	15, 768	1, 386, 908	....

ILLINOIS.

(No official returns of sugar.)

Years.	Acres.	Sirup.
		Gallons.
1879	17, 883	1, 309, 400
1880	9, 825	636, 216

KANSAS.

Years.	Acres.	Sirup.	Sugar.
		Gallons.	*Pounds.*
1872			
1873			
1874	14,103		540,338
1875	23,026		1,149,030
1876	15,714		839,147
1877	20,784	2,390,131	1,195,066
1878	20,292	2,333,566	1,166,783
1879	23,665		1,224,557
1880			

For twenty-five years past the average yield of sirup, varying from 16,000,000 gallons per annum to 5,000,000 or 6,000,000, has probably averaged about 11,000,000 gallons, valued at 65 cents to 40 cents. For sirup, fodder, and all purposes, the average value of the crop may have approximated $8,000,000 per annum.

J. R. DODGE,
Statistician.

6.—SORGHUM SUGAR-CANE.

NEW JERSEY AGRICULTURAL EXPERIMENT STATION.

SORGHUM SUGAR-CANE.—EXPERIMENTS ON ITS GROWTH AND SUGAR PRODUCT.

For the last forty years there have been experiments made to manufacture sugar from maize and from sorghum, and during the late civil war sorghum was grown in large quantity for the production of sirup, especially in the Western States. More recently the United States Commissioner of Agriculture has given much attention to the growing of sorghum, and to the manufacture of sugar from it. His reports for 1879 and 1880 contain much interesting and valuable matter upon the subject. The growing of sorghum and the manufacture of sirup from it has come to be an established branch of farm industry in several of the Western States, and as more knowledge and skill are acquired, it is found that good granulated cane sugar can be made from it in paying quantities. A large amount of both sirup and sugar are now made every year in Kansas and adjoining States, and the proof appears to be complete that with a proper establishment for the manufacture, and skillful workmen, the business can be profitably carried on here.

During the last session of our legislature a bill was passed entitled "An act to encourage the manufacture of sugar in the State of New Jersey." This act provides that a bounty of $1 per ton may be paid by the State to the farmer for each ton of material out of which crystallized cane sugar has actually been obtained; it provides also a further bounty of one cent per pound to be paid to the manufacturer for each pound of cane sugar made from such materials. After the passage of this act, the Senate, on motion of Senator Taylor, requested the Agricultural College to experiment on the sorghum plant, in order to further its cultivation by the farmers of this State. The following bulletin is published in compliance with this request.

The sorghum was grown on the college farm, and the chemical work carried out in the laboratory of the experiment station. The investigation includes the trial of different varieties of sorghum with special reference to their time of ripening and percentage of sugar, as well as the study of the effect of different fertilizing ingredients, applied singly and in combination, upon the yield of sugar and seed.

The field selected for the experiment is thoroughly underdrained, rather heavy piece of land, cropped last year with field corn grown on sod, to which a liberal dressing of barnyard manure had been applied. On that portion devoted to the trial of different varieties, Mapes's sorghum manure was used this year immediately before planting, at the rate of 600 pounds per acre. The seeds were kindly furnished by Dr. Peter Collier, chemist of the United States Department of Agriculture.

Dr. Collier in his valuable reports has clearly shown that the condition of the ripening seed may be taken as an index to the condition of the juice of the plant. When the seeds have become so hard that they can no longer be split with the fingernail the stalks will contain the maximum amount of sugar and minimum of glucose, and when this stage is reached the plant may be regarded as matured.

The importance of using great care in the choice of seed is illustrated by the following list of varieties:

Wolf Tail..	Failed to mature before frost.
Link's Hybrid	Do.
Liberian..	Do.
Early Amber	Seed failed to germinate.
Neeazana	Failed to mature before frost.
Goose Neck	Matured.
Sorghum...	Do.
Early Orange	Failed to mature before frost.
Oomeeseana	Matured.
Gray Top...	Failed to mature before frost.
African	Matured.
Honduras	Failed to mature before frost.
Chinese...	Do.
Early Golden......................................	Matured.

Of the fourteen varieties five only matured. Their relative value to the manufacturer is shown below:

	Goose Neck.	Sorghum.	Oomeeseana.	African.	Early Golden.
Percentage of juice.....................................	60.3	61.4	58.8	57.5	60.0
Percentage of sugar in juice...........................	8.58	7.28	6.50	7.60	14.06
Pounds of extractable sugar per ton	104	89	76	87	169

In a table on the following page it will be noticed that the Early Amber also matured; yielding under favorable conditions 162 pounds of sugar per ton of stripped and topped cane. Judging from this experiment the choice of variety for this section of the State is limited to the *Early Amber* and *Early Golden*.

For the study of the effect of fertilizers sixteen adjoining plots, of one tenth acre each, were measured off, fertilized as stated in the table, and planted May 23, 1881, with Early Amber seed. The cane was doubtless injured by the unusually severe drought; it was noticeable, however, that it suffered much less from this than corn planted on a neighboring field. It was harvested on the first of October.

For samples to represent each plot, twenty average canes were cut from ten different rows, immediately, weighed and after they had been stripped and topped again weighed and passed singly between the rollers of a heavy cane mill. The juice from each lot of twenty cane after it had been carefully mixed was used for the analysis. The determinations of cane sugar were made by means of the polariscope, using solutions clarified with basic lead acetate and 50 per cent. absolute alcohol.

The plan of the experiment was, to ascertain the effect of each of the fertilizing materials applied singly and in combination on the production of sugar—to compare the effect of muriate of potash with that of sulphate of potash—and to determine whether by increasing the amount of phosphoric acid used per acre an advantage would be gained. The action of the fertilizers is best studied in the table under the heading, pounds of extractable sugar per ton of cane and per acre. It was expected that phosphoric acid would materially hasten the maturity of the cane; it appears to have exercised no decided influence in this respect. It caused, however, an increase of 250 pounds or nearly 30 per cent. of sugar over that yielded by plot No. 15.

Muriate of potash used alone increases the gross weight of stalks very much more than sulphate of potash; it increases too the yield of sugar per acre. It is a fact, however, of especial importance to the manufacturer that the yield per ton is 20 per cent. greater from the plot No. 12, on which the sulphate was used, than from the muriate plot No. 4. Muriates, too, if taken into the sorghum juice, cannot be removed by the process of manufacture now used, and interfere seriously with the crystallization of sugar.

As has been well known for many years past, crude barn-yard manure must not be used directly on sugar-producing plants. Plot No. 11 draws attention once more to the fact. No noticeable increase in the amount of sugar was caused by it; but a point of much greater importance is the positive statement of experienced men that sugar will not crystallize from sirup of canes which have been fertilized with it. A heavy dressing on corn land loses its injurious qualities in the course of a year, and sorghum following corn in rotation is benefited by it.

The expression "extractable sugar" has been used in this table to indicate tha a portion only of the total amount of sugar has been extracted by the mill; the bagasse or crushed cane when it is burned under the boilers or thrown on the compost heap still contains one-third of the sugar produced by the plant. If the profits of the business are so large that manufacturers can content themselves with two-thirds of the sugar, farmers should endeavor to turn this bagasse into food for sheep, by the process of ensilage. After a struggle which has now lasted more than twenty-five years, sorghum to-day does not occupy its true position among sugar-producing plants. Its advocates justly claim that this is due to our lack of information, not only in regard to the manufacture of sugar from it, but also in respect to its proper cultivation.

	No manure.	350 pounds superphosphate.	150 pounds nitrate of soda.*	150 pounds potassium chloride.	350 pounds superphosphate, 150 pounds sodium nitrate.	No manure.	350 pounds superphosphate, 150 pounds potassium chloride,	150 pounds sodium nitrate, 150 pounds potassium chloride.
Cost of fertilizers per acre	$0 00	$6 10	$7 50	$3 40	$13 60	$0 00	$9 50	$11 00
Pounds of sorghum per acre	11,515	13,345	14,820	16,000	14,440	11,170	11,640	12,390
Pounds of stripped and topped cane per acre	8,406	9,890	11,263	12,160	10,830	8,378	8,846	9,298
Per cent. of juice extracted from stripped and topped cane	69.6	67.0	66.4	68.0	66.8	64.2	65.1	64.3
Pounds of juice extracted per acre	5,851	6,626	7,479	8,209	7,234	5,379	5,759	5,975
Per cent. of sugar in juice	9.70	9.43	9.06	9.27	9.68	9.94	10.51	11.65
Pounds of extractable sugar per acre	568	625	673	767	700	535	605	696
Pounds of sugar extracted per ton of cane	135	126	120	126	129	128	137	150
Pounds of clean seed per acre	1,020	1,351	1,298	1,246	1,344	1,132	1,038	1,067

	350 pounds superphosphate, 150 pounds potassium chloride, 150 pounds sodium nitrate.	400 pounds calcium sulphate.	20 two-horse loads barn-yard manure.	200 pounds potassium sulphate.	350 pounds superphosphate, 200 pounds potassium sulphate.	200 pounds potassium sulphate, 150 pounds sodium nitrate.	200 pounds potassium sulphate, 150 pounds sodium nitrate, 350 pounds superphosphate.	200 pounds potassium sulphate, 150 pounds sodium nitrate, 700 pounds superphosphate.
Cost of fertilizers per acre	$17 10	$1 60	$40 00	$6 50	$12 60	$14 00	$20 10	$26 20
Pounds of sorghum per acre	12,590	12,680	12,375	11,605	11,650	12,505	13,260	14,030
Pounds of stripped and topped cane per acre	9,946	9,510	9,405	8,820	8,854	9,504	9,812	10,943
Per cent. of juice extracted from stripped and topped cane	67.3	64.8	64.4	68.2	69.0	68.3	68.9	67.8
Pounds of juice extracted per acre	6,694	6,162	6,057	6,015	6,109	6,491	6,760	7,419
Per cent. of sugar in juice	11.43	9.84	9.57	11.61	11.73	9.73	9.44	12.01
Pounds of extractable sugar per acre	765	606	580	698	594	632	638	891
Pounds of sugar extracted per ton of cane	154	128	122	158	134	133	130	163
Pounds of clean seed per acre	1,305	1,136	1,160	1,216	1,226	1,139	1,349	1,278

*16 per cent. phosphoric acid.

For some time past authorities have felt that the hope of having a small sugar-house on each farm must be abandoned and that our attention must be turned towards the more rational plan of thoroughly-equipped manufactories in which the sorghum grown on neighboring farms can be worked quickly and economically by skilled operatives. This plan is now on trial at Rio Grande, Cape May County. Mr. Hilgert, an enterprising and energetic business man of Philadelphia, member of the firm J. Hilgert's Sons, sugar refiners, has built and fitted up an extensive sugar-house at an expense of at least $60,000. This house during the past fall worked the cane of about 700 acres. The product of crystallized sugar was sold to refiners at 7 and 8 cents per pound. The yield, although not as large as expected, is still regarded as satisfactory. The farmers of that section who calculated on an average yield of 10 tons of cane and 30 bushels of seed per acre have been disappointed, the average yield per acre having been about 5 tons of cane and 20 bushels of seed, which sold readily for 65 cents per bushel. Mr. Miller, who was perhaps the largest cane-grower on the cape, raised on 120 acres 641 tons of cane and 2,500 bushels of seed. The total amount realized by him is reported to be $3,648. The cost of growing this crop is not known at present, but the reported cost for Iowa in the year 1873 is, exclusive of fertilizers, $12.50 per acre.

The result of the season's experiments is decidedly encouraging, considering the unfavorable circumstances. There has been a drought of unprecedented severity and length, so that the corn crop on the college farm was not more than one-quarter its usual amount. And yet the results of sorghum-growing on the same farm, as given in the above table, are respectable. With a season having the average rainfall a crop weighing from two to three times as much as that of the present one may safely be calculated on.

The expense of hoeing and cultivating sorghum in the earlier stages of its growth are much greater than for field corn, and to those only accustomed to growing the latter crop it is discouraging. The plants are very small when they first come up, and look so much like common summer grasses that they may be mistaken for them, and for several weeks the grasses and weeds grow much the fastest. The later stages of growth of the sorghum are very rapid. Those who intend to grow sorghum must then be very watchful of it in the early part of the season. It is most commonly planted in drills from 3 to 3½ feet apart, with hills about 15 to 18 inches apart and having six or seven seeds to a hill. Some, however, plant it in hills with rows 3 feet apart both ways. Each method has its advocates, but the latter costs the least for labor, and advocates for the other method claim that it does not yield nearly as much per acre. There is much to be learned in this respect by our farmers, and experiments should be made with care.

The soil best adapted to it is said to be a sandy loam, though it will grow well on any ground that will produce Indian corn. It grows well on the same field year after year, only care being taken to keep the field rid of the seeds of weeds. A manure containing large percentages of sulphate of potash, a soluble phosphate of lime, and not much ammonia is probably the best and most economical for its growth.

The value of the crop is considered to be mainly in the sugar, but the seed is found to be about equal to Indian corn in feeding value, and the crop per acre is not less than that of other common cereals. There are no good feeding experiments to show what may be the value of stalks from which the juice has been extracted.

The field for enterprise in this direction is a large and inviting one, and it is to be hoped that the promise of the manufacture, and the bounty offered by the State, may lead to the permanent and profitable establishment of this branch of industry in our State.

GEO. H. COOK, *Director.*

NEW BRUNSWICK, N. J., *December 20, 1881.*

7.—*RIO GRANDE SUGAR COMPANY, NEW JERSEY.*

A.—*Letter from the president of the Rio Grande Sugar Company to the Tariff Commission.*

SORGHUM SUGAR.

OFFICE OF THE RIO GRANDE SUGAR COMPANY,
Rio Grande, N. J., September 29, 1882.

DEAR SIR: I have the honor to submit to you the following statement touching the production of sorghum, and the manufacture of sugar and sirups therefrom:

It will be needless to refer you to the reports on this subject of the Department of Agriculture, as these, doubtless, have been already placed before you.

Thirty years ago the raising of sorghum sugar-cane created quite an excitement in this country, owing to the promised revolution it was to effect in the sugar production of the country. The failures that ensued are well known, and the production of sugar from northern cane has only had spurts of success, and up to about the present time has resulted only in disaster.

The raising of sorghum cane, now practiced among small farmers, produces only the crude molasses for which local demand may exist.

During these thirty years the cane has been continually improving in quality, and yielding largely increased amounts of sugar in the juice.

The last report of the Department of Agriculture names some producers who returned nearly 16 per cent. of sugar. We have found it profitable to work it when as low as 8 per cent.

Of first importance in the raising of sorghum cane is the selection of a proper climate.

Second. A proper soil.

Third. Skillful fertilization.

Fourth. Proper appliances for conducting the processes in a systematic and skillful manner.

In the report I am about to give you of the incipiency and success of the Rio Grande Sugar Company these matters have been carefully looked after.

With regard to the selection of the land, it was made with a view of being near the waters of the ocean and bay, as the settlement of "Rio Grande" is about 4 miles distant from, and between, the Atlantic Ocean and Delaware Bay. At such points the early frost of our autumn does not reach an injurious effect within thirty days of the time it does in the inland country but a few miles distant.

The season for growing the cane is comparatively a short one, consequently a great gain arises from the longer period of time that can be secured for the late production of cane, and when it may be given ample time to come to a ripened condition.

While upon this point it is well to note the large area of country that is applicable to this culture, and that fulfills the requisites as I have stated them; for example, the large area included in the peninsula of which we are now speaking, as well as of the Delaware and Maryland peninsula, lying between Chesapeake and Delaware Bays.

In raising sorghum in such a limited period of time for its growth the soil is a highly important element.

The character of that occupied by the Rio Grande Company has demonstrated the fact that a rich soil is not a requisite; it can be said that a comparatively poor and sandy soil may be considered the most desirable for this purpose, as the plant cannot absorb or receive any large quantities of salts (say potash salts) from such soil.

The plant of these works has been erected in the most approved fashion, and in size and completeness will compare favorably with those of Cuba and the State of Louisiana.

The Rio Grande Sugar Company was organized in the year 1881 with a capital of $250,000, based upon the works being almost completed at that time by private hands, when an additional capital was taken, and the lands purchased, as it was shown that the only safe method of raising sorghum was for the company to undertake it, as the farmers immediately surrounding the locality, although able to produce it of proper quality, were not to be entirely depended on to deliver it in good condition, and from fields in the order of fitness for grinding.

The drought of 1881, together with the fact just mentioned, made the raising of cane during that season of small account.

In the spring of 1882 about 1,000 acres of the 2,400 owned by the company were planted and liberally manured with Peruvian guano, enriched with sulphate of ammonia.

The cane planted was chiefly of the Amber variety, as this ripens in three months after planting.

Sufficient demonstration now exists to prove that sorghum can be successfully and profitably raised in the manner already mentioned.

In the year 1880 the State of New Jersey passed a law granting a bounty of 1 cent a pound on sugar, and $1 a ton on cane produced for the term of five years; therefore but three years now remain during which it may be taken advantage of. However, without this bounty, and with the existing rates of import duty on sugar, it is altogether possible that within the next decade the regions spoken of will be largely occupied by planters of sorghum, as these lands require some change of crops to relieve them from a condition of poverty.

A large portion of Virginia and North Carolina will ultimately produce this cane, and it can be said that the benefit derivable from showing the way to improving such a large area of land is one of the most important considerations now existing in the United States.

That the country should, in a short prospective period of time, be on the highway to independence in her sugar production, will add also to the interesting features of this matter.

The Rio Grande Sugar Company was established in view of continued protection by the Government of the United States. I trust, therefore, it will not be a part of the acts of the Tariff Commission to reduce the duties on sugar, and thus possibly extinguish in its infancy so important a branch of industry.

Very respectfully, your obedient servant,

GEORGE C. POTTS,
President.

Hon. JOHN L. HAYES,
President Tariff Commission.

B.—*Capt. R. Blakeley describing his visit to the Rio Grande Sugar Works.*

SAINT PAUL, *September* 29, 1882.

DEAR SIR: On my way from your place to Washington I called at the office of Mr. G. C. Potts, at Philadelphia, and presented your letter. Mr. Potts was down at the mill, but returned in the evening and called upon me at the hotel and cordially invited me to go down and see their works, which I did, and was especially gratified with my reception by Mr. Hughs, the chemist in charge of the mill, who took great pleasure in showing me over the mill and farm. This place is situated about 5 miles north of Cape May, upon a sandy soil, about 8 feet above the salt water. The cane (about 900 acres) is well grown, except the part which was replanted. The seed did not all come up, and they attempted to plant more, and happened to get a good growth by that means; but it failed to mature, and has, consequently, injured the balance of the crop; but still it is the best crop that has ever been grown, so far as I know. They used about $6,000 worth of guano upon the crop, as Mr. Hughs informed me, but this land must be fed or it will not do for cane. Mr. Hughs says they will have 9,000 tons of cane and 20,000 bushels of seed. They sold the seed last season for 65 cents per bushel, for feed, and have parties who stand ready to take the whole crop this season, but, possibly, at a reduced price, as corn is likely to be lower, and the seed will not be as high as last season, but they are promised a corresponding price for this season.

Their mill is put up without regard to expense, although not in just as good shape as they would build it if they had it to do again. The machinery was built by Morris, of Philadelphia, and is very substantial; rollers 5 feet long and 30 inches in diameter; engine 125 horse-power, and boiler surface is equal to 600 horse-power; four defecating pans of 500 gallons each; four open evaporators, with copper pipe for the steam, and two vacuum pans to complete the evaporation, and all work very well, except the vacuum pans. These are not equal to the other parts of the mill, as they will not take care of more than the produce of 150 tons of cane in twenty-four hours, while the mill would work 300 tons of cane in same time. They also have one filter-press for the scums, by which they save 1,000 gallons of juice in twenty-four hours, which is usually wasted.

The men who are doing the work in the mill are, most of them, skillful men, especially the men who are doing the boiling and finishing work of the manufacturing of sugar. They are making sugar for the refineries, and get 7 cents per pound at Philadelphia, and are not using any bone coal at the factory. The sugar is run into wagons from the strike pans, and stands about three days to crystallize, and is then run through the mixer to the centrifugals, and barreled and sent to market.

Mr. Hughs says they are averaging 8,000 pounds a day of sugar. They are only taking out the first sugar now, as they have not time to reboil; but the sirup from the centrifugal is barreled to wait until after the grinding season is over.

There are different kinds of cane grown in this place, but the dependence is upon the "Amber Cane." They have a very nice piece of "Early Orange" that promises well. Their crop polarizes from 9 to 15 per cent.

Mr. Hughs is perfectly enthusiastic in regard to the future of this business. The company have 2,000 acres of land, and he thinks they will plant nearly the whole of it next season and will put in another vacuum pan, so that they can handle all the juice from 300 tons of cane per day.

The cane is, when fully ripe, or mostly so, cut and seed cut off and piled in the field to dry. The cane with the leaves on is brought to the mill and ground day and night, so that in the morning the cane that was cut the day before has all been ground and worked continuously until finished.

Mr. Hughs says that this crop will produce about 80 pounds of sugar to the ton, if no accident happens before it is all finished up. This is a very strong confirmation of all the hopes and expectations of the most sanguine friends of the sorghum-sugar industry.

On my arrival at Washington I called upon Professor Collier and handed him a sample of sugar which Mr. Potts sent from their works, and informed him of what I had seen and expressed the hope that he would be able to visit the mill at Rio Grande during the first week in October, and would have asked the Commissioner to do so, but he was absent from the city. I hope that when he returns he will finally accept the invitation of the Rio Grande Company and visit their works.

I hope that you may be able to go down next week and see for yourself.

Respectfully,

                                       R. BLAKELEY.

Prof. B. SILLIMAN,
    New Haven, Conn.

C.—Mr. Harry W. McCall, of Donaldsonville, La.: A sugar planter's views of the Rio Grande Sugar Company's plantation, in a letter to the chairman.

ALBEMARLE HOTEL, NEW YORK, *October* 2.

DEAR SIR: Your letter of September 30 has just reached me here. I remember, when visiting the Rio Grande sorghum plantation, that Mr. Potts stated that you were coming there, and it is quite true that I was quite favorably impressed by what I saw.

In regard to the special questions which you ask in your letter—whether it appears to me that the sorghum merits "the character of a sugar-producing plant, adapted to a wide area of land outside of existing sugar districts"; also, if strong varieties of it might be cultivated "with success in competition with Ribbon cane"—I would say, as to the first one, that I certainly think it merits the character of a sugar-producing plant, and that it is likely it will be found adapted to wide areas outside of the existing sugar belt. Like the Blue and Ribbon cane of Louisiana, I understand that it is killed or injured by a freeze, so that most of it must be gathered before this occurs. But it has this great advantage over our southern cane, that it may be planted late—in April say—and matures sufficiently to allow of cutting early in September, and, perhaps, earlier in some varieties. This fact, together with another, that it is propagated from seed instead of eyes, renders it, in my opinion, adapted to cultivation over a very wide extent of country.

Your other question, whether I see reason to believe it could compete successfully with the Ribbon cane of Louisiana, I must answer by saying, as yet I do not, though it may, perhaps, turn out so. As I do not think there are data enough yet to answer this question decidedly, I have not formed an opinion on it. But I see no reason why any results obtained from sorghum at the North should not be also obtained at the South. So if it could be clearly proved a profitable crop at the North, it would be well worth trying at the South. But in order to do this we must have full figures and be sure they have been accurately kept.

At the time I made my visit to the Rio Grande, they had just begun, and had not had time to get their figures together, as I hope they will do before they finish. I saw enough to make me believe that the undertaking is a success. But to judge properly we must know how much sugar they get to the ton of cane, and how many tons to the acre, and what the total cost of making and cultivation, which they could not know with any certainty at the time I was there, though they appeared to be doing so well that I thought Mr. Potts quite justified in anticipating a profit. I can say no more for the present.

If any of the above remarks prove of interest to you, you are most welcome to them, as also to any other information I may be able to give you in the future.

Very respectfully, yours,

HARRY W. McCALL.

To B. SILLIMAN, Esq.,
   New Haven, Conn.

D.—Rio Grande Sugar Company's form of returns required by the State of New Jersey to secure bounty.

NEW JERSEY, *ss* :

——— ——— ———, being duly sworn, saith that the following is a correct and just statement of the quantity of ———, grown by him in the county of ———, for the purpose of being manufactured into sugar in this State, and that the same has been delivered at the manufactory of the sugar refinery, at Rio Grande Station, Cape May County, New Jersey.

	Dollars.	Cents.
Beets	——	—
Sorghum	——	—
Sugar-cane	——	—

Sworn and subscribed the ——— day of ———, 188–, before me.

—— ———.

I, ——— ———, clerk of the county of ———, do hereby certify that the aforesaid ——— ——— is a resident in good repute of the county of ———, in the State of New Jersey.

——— ———, *Clerk.*

Manufacturer's certificate.

I hereby certify that ——— ——— has delivered at the manufactory of the Rio Grand Sugar Company, at Rio Grande Station, Cape May County, ——— pounds of

——, stated to have been grown by him, and that the same has been manufactured into sugar.

Dated ——— ———, 188-.

I do hereby certify that the above account is correct.

——— ———,
Chief of the Bureau of Statistics of Labor and Industries.

8.—CHAMPAIGN SUGAR AND GLUCOSE MANUFACTURING COMPANY CHAMPAIGN, ILLINOIS.

The following report of the operations of this company for the season of 1882 brings down the results only to October 28, being still some time before the completion of the rolling of cane. The results speak for themselves:

CHAMPAIGN, ILL., *October* 28, 1882.

SIR: The undersigned have the honor to present to you the following report on the manufacture of sorghum sugar for the year 1882. Our report is necessarily incomplete, as we are still in the midst of our season's work. But the gratifying results thus far obtained will, we hope, warrant reporting the data at hand.

HENRY A. WEBER.
MELVILLE A. SCOVELL.

Prof. B. SILLIMAN,
Chairman of Sorghum Sugar Committee, National Academy of Science.

As a result of the experiments carried on by the writers in the seasons of 1880 and 1881, the Champaign Sugar and Glucose Company, of Champaign, Ill., was organized. The object of the company was to carry out on a commercial scale the production of sugar and glucose from sorghum, as was indicated by our laboratory experiments. The company was organized with a capital stock of $25,000. The total expenditure for building the works and raising the crop, however, exceeds $30,000. The main building is 40 by 60 feet and three stories high, with a "lean-to," 45 by 30 feet, covering the engine and crushers. Near the main building are situated the boiler-house, with two ninety horse-power boilers, and a kiln with twelve retorts for revivifying the bone-black.

For the sake of convenience the description of the apparatus will be given in connection with the process followed in manufacturing sugar and sirup.

The cane is conveyed by means of a carrier 50 feet in length to the first mill, a "Cuba No. 4," manufactured by George L. Squire, of Buffalo, N. Y., who kindly consented to the use of his rubber springs for our second mill, which was originally one of the rigid kind.

After leaving the first mill the bagasse is moistened with a spray of hot water and is conveyed, by means of an intervening apron, to the second mill. By the use of this second mill the sugar which is left in the bagasse, after passing through a single mill, as is pointed out in the report of our experiments, is practically all recovered.

The juice from the two mills is pumped together to the juice tanks, which are placed at the top of the main building and have a capacity of about 3,000 gallons. From here it is drawn to the defecators, where it is exactly neutralized with milk of lime in the cold, heated to the boiling point and thoroughly skimmed. These defecators are made of wood lined with galvanized iron and supplied with copper coils for heating. Four of them have a capacity of 660 gallons each, and one of over 1,300 gallons. After settling, the juice is allowed to run into the evaporators, where it is concentrated to a density of 25° Beaumé. The evaporators are two in number, 8 feet in diameter, made of copper, and supplied with copper coils. From the evaporators the liquor runs into settling tanks, and next through bone-coal filters. The filters are four in number, 2 feet in diameter, and 12 feet high. The liquor is next drawn up into the vacuum pan, where it is concentrated into melada. The crystallization of the sugar takes place in the vacuum pan, and could at once be run into the mixer and centrifugals. Owing to the fact that only one centrifugal has thus far been supplied, the strikes from the pan are usually run into crystallizing wagons, and placed in a warm room until the sugar can be "swung out." There are fifty of these wagons having a capacity of 120 gallons each.

The quality of the sugar produced is unobjectionable in regard to taste and color. It grades as extra yellow C, and sells readily at the factory at 8¼ cents per pound in lots of five barrels. The molasses is of a dark color, but still rich in cane sugar. It is stored up in barrels and will be kept until the cane is all harvested, when it will either be refined or worked over for a second yield of sugar.

The company raised 190 acres of cane, 8 acres of which is "Kansas Orange," about 40 acres "Early Orange," and the rest "Early Amber." Private parties planted about 100 acres more, all of which was Early Amber, with the exception of one field of Early Orange containing 12½ acres.

The company began working up their Amber cane on September 21. An analysis of the juice was made, with the following result:

Specific gravity... 14. 8° Brix.
Cane sugar .... 8. 20 per cent.
Grape sugar .... 3. 63 per cent.

The best Amber cane of the company was grown on sod ground, the field containing 50 acres.

The composition of the juice of this field on October 21 was as follows:

Specific gravity... 1. 060
Cane sugar .... 10. 17 per cent.
Grape sugar .... 2. 48 per cent.

Owing to the lateness of the season one continuous run was made, and the cane raised by private parties was worked up with the company's cane, so that it will be impossible to give the yield per ton and acre before the close of the season's work.

One field of Early Orange, grown by Mr. J. G. Clark, has been harvested by itself and the products kept by itself. Of this field and variety of cane exact data can be given. The composition of juice October 24 was as follows:

Specific gravity .... 1. 070=16. 3° brix.
Cane sugar .... 10. 82 per cent.
Grape sugar... 3. 54 per cent.
Number of acres in field 12. 5
Total amount of cane stripped and topped 156   tons.
Yield per acre... 12. 5 tons.
Amount of juice.. 20, 939 gallons.
Weight of juice .... 185, 947 pounds.
Per cent. of juice.. 59. 6
Weight of melada .... 25, 920 pounds.
Weight of sugar .... 9, 900 pounds.
Weight of molasses .... 16, 020 pounds.
Gallons of molasses.. 1, 456
Yield of sugar per acre... 790   pounds.
Yield of molasses per acre.. 116. 5 gallons.

In this statement the amount of water added in moistening the bagasse before passing through the second mill has been deducted from the total amount of juice obtained.

The melada obtained from the Amber cane is fully as rich in sugar as that obtained from the Orange. The yield of sugar and molasses per acre will be lower for some of the fields of Amber, but for others it will be fully as high and in a few cases perhaps even higher.

It is not more than fair to add that for this section of the country the season has been very unpropitious for the proper development of sorghum cane. This will be seen at a glance by comparing the analyses given here with those made in this locality last year and the year before, as given in our report. The necessary hot summer temperature for the production of a high percentage of sugar was entirely wanting. But, on the whole, the sorghum-sugar industry is to be congratulated for this cold, wet season, as the flattering results, which we are nevertheless obtaining here, will forever silence the claim that sugar can be made from sorghum only under the most favorable circumstances.

9—EXPERIMENTS IN AMBER CANE, ETC., AT THE EXPERIMENTAL FARM MADISON, WIS., 1881.

REPORT.

To his Excellency J. M. Rusk, *Governor:*

In conformity to chapter 211 of the general laws of 1881, I herewith present a report of the experiments in Amber cane and ensilage of fodders, conducted upon the university experimental farm the past season.

Most fortunately, Mr. Magnus Swenson was secured as chemist in these experiments, and too much credit cannot be given him for his untiring zeal in the difficult task to which he was assigned. Such an experiment as securing sugar from Amber cane in anything like a practical way is a most difficult undertaking. Every step in the process is along an unknown road, and the many failures in past years show that scores of persons who thought they were certain of success only attained defeat.

Fortunately Mr. Swenson understands machinery as well as chemistry, and was enabled to design and superintend the construction of the machinery used. By this means a great saving was effected in the cost of machinery needed. Had it been otherwise, the funds would not have been sufficient for the work.

I present Mr. Swenson's report as handed to me, believing that in it those interested in Amber cane will find information that cannot but prove of great value to them. The fact that good marketable sugar can be obtained from Amber-cane at the rate of 1,000 pounds to the acre, by methods even more practicable when used on a large scale than in the present case, is a cause for gratification, I think.

It is proposed to distribute samples of sirup and sugar obtained in the experiments, in such a way that they can be seen at all the agricultural gatherings held this winter throughout the State.

Having experimented but a single season, it is needless to say that much remains to be done yet, and many problems are still awaiting solution.

In addition to the experiments, I have tried to learn the condition of the industry throughout the State and have taken steps to familiarize our farmers with what we are trying to do.

In April last a twelve-page circular relative to Amber cane was prepared and 3,000 copies distributed.

This fall 1,500 copies of a circular letter, making inquiries regarding the cane crop, were prepared and sent to all whom I thought could aid us. In answer to these circulars I have replies from 180 manufacturers of Amber-cane sirup, who report having made about 350,000 gallons of sirup this year. A list of these manufacturers, together with amount of sirup made by each, is herewith given. Other valuable information from these reports is given in its proper place.

In regard to the second experiment, the ensilage of fodders, permit me to say that a silo was built and filled last summer, and experiments are now in progress to determine the value of the ensilage. So far the indications are very favorable, but it is too soon to make any definite statements. As complete a report as possible is herewith presented. It is planned that Mr. Swenson investigate the subject, from the chemical side, this winter, and upon this point much remains yet to be known.

As required by the act above named, I have made a detailed statement of the moneys expended up to the present. It will be seen that we have not yet expended the sum granted.

Most respectfully submitted.

W. A. HENRY,
Professor of Agriculture, University of Wisconsin.

EXPERIMENTAL FARM, UNIVERSITY OF WISCONSIN,
Madison, Wis., December 31, 1881.

EXPERIMENTS WITH SORGHUM CANES.

BY MAGNUS SWENSON.

The chief object of the experiments conducted during the past season (1881) has been to demonstrate the practicability of making sugar from cane grown in this State. For this reason the work has been carried on in a thoroughly practical manner. My results are not based on theory; they do not show what might be obtained, but what has actually been done. The amount of sugar obtained is not deduced from the amount present in the cane or sirup, but represents what has actually been crystallized and separated as sugar.

MACHINERY.

The apparatus used consisted of one horizontal mill, made by the Madison Manufacturing Company; one ten horse-power steam boiler; one defecator of galvanized sheet iron, 3 feet high, 2.5 feet in diameter, and heated by a steam coil made of 1-inch gas pipe; two galvanized iron evaporating pans, the larger 6 feet long, 3 feet wide, 1 foot deep; the smaller 4 feet long, 2 feet wide, 8 inches deep, both heated by steam coils; one globular vacuum pan 30 inches in diameter; one wet air-pump for exhausting the vacuum pan; one centrifugal machine for separating the sugar from the sirup, 1¼ feet in diameter and 4 inches deep; one small steam pump for feeding the boiler and running the vacuum pan and centrifugal machine.

CANE-SUGAR AND GLUCOSE.

Before passing on to the actual experiments, a few pages will be devoted to the general properties of cane-sugar, and the substances occurring with it in the cane juice. The average cane contains about 85 per cent. of juice and 15 per cent of dry bagasse. The juice from the average cane obtained on the farm consisted of 9.5 per cent. cane-sugar, 3.2 per cent. glucose, 2.3 per cent. organic acid and vegetable matter, and 85 per cent. water. Cane-sugar is a compound substance composed of 12 parts carbon, 22 parts hydrogen, 11 parts oxygen ; or, since 1 part oxygen and 2 parts hydrogen form water, we may consider cane-sugar to be made up of 12 parts carbon and 11 parts water.

Glucose, or grape sugar, as it is also called, is composed of 12 parts carbon, 24 parts of hydrogen, 12 parts of oxygen, or 12 parts carbon and 12 parts water. The only difference between the two is 1 part of water. If a solution of cane-sugar in water is heated with a small quantity of almost any acid, it takes up one more part of water, and thus becomes changed to glucose. Almost the same thing takes place when a solution of cane-sugar is acted upon by a ferment, such as yeast, or, even by simply heating for some time, large quantities of the crystallizable cane-sugar are changed. The one important thing in the boiling down of cane juice is to guard against this change. As seen before, the destruction of cane-sugar may be induced in three different ways: 1st. By the presence of an acid. 2d. By the presence of a ferment. 3d. By high and prolonged heat. We will discuss them in order.

PRESENCE OF AN ACID.

All cane juice contains a considerable proportion of free organic acids. If, therefore, the juice be boiled down without first neutralizing these acids, a large part of the cane-sugar will be changed into glucose. The amount of cane-sugar destroyed may be seen from the following experiment: Six hundred pounds juice, containing 9.96 per cent. cane-sugar and 3.45 per cent. glucose, was taken directly from the mill and boiled down to sirup. The sirup was found to contain 22.4 per cent. cane-sugar and 56.3 per cent. glucose. If no inversion had taken place, the sirup should have contained 58.3 per cent. cane-sugar; so we see that 61.6 per cent. of all the cane-sugar originally in the juice had been changed into glucose. Glucose has only one-third the sweetening power of cane-sugar, and its presence prevents, to a large extent, the crystallization of cane-sugar. The light-colored, putty-like deposit in Amber sirup, which is often mistaken for cane-sugar, is glucose.

USE OF LIME.

If lime is added to the juice it will combine with and neutralize the acid, and this union of the lime and acid forms a new substance, which becomes, to a large extent, insoluble, and is removed with the scum, what remains in the solution having no effect whatever on the cane-sugar. But here we meet with another difficulty. If more lime than is necessary to neutralize the acid has been added, although the excess has no effect whatever on the cane-sugar, it will at once begin to decompose the glucose, changing it into a series of very dark and bitter products, which will impart a dark color, and a bitter, burnt taste to the sirup. Fortunately we are in the possession of a very simple test which tells when lime enough has been added. If a piece of blue litmus paper is dipped into water containing a small quantity of acid, it at once turns red ; and if a piece of red litmus paper is dipped into water made slightly alkaline by the addition of a little lime water, it at once turns blue. If, now, to a portion of the acidified water we add gradually some lime water, we will soon arrive at a point when the solution will have no effect on the color of either red or blue litmus; in other words, it is neither alkaline nor acid, but neutral. This will be treated of again under the head of defecation.

FERMENTATION.

The next thing which tends to destroy the cane sugar is fermentation. This process begins almost immediately after the juice leaves the mill, and when the weather is warm large quantities of sugar are lost in this way. Fermentation is at once arrested by heating the juice to near the boiling point. Cane juice should therefore never be allowed to remain standing any length of time, but should be defecated as soon as possible after coming from the mill.

HIGH TEMPERATURE.

High and prolonged heat is very destructive to crystallizable cane sugar. At first the temperature will not vary much from that of boiling water, or 212° F., but as it becomes more and more concentrated the boiling point gradually rises, until, when the sirup is thick enough for sugar making, the boiling point is from 232° to 234°. The destruction of sugar takes place long before this point is reached. To get the best results the sirup should not be boiled in an open pan after it reaches a density of 20° B., but should then be transferred to the vacuum pan. During the first part of the boiling in this pan the temperature should not exceed 170° F., and when the sirup becomes denser a more complete vacuum should be maintained so as to boil it about 140° F.; in fact, the lower the temperature the better.

The varieties of cane raised on the farm during the past season were confined to the Early Amber, Early Orange, and Honduras. Of these the Early Amber is unquestionably the best for sugar making, and our experiments were confined largely to this variety. The total amount of juice in this cane is about 85 per cent. of the total weight of the stalks, and the juice contained 9.20 per cent. cane sugar and 3.4 per cent. glucose. This content of sugar represents the average of not less than 200 pounds of stalks stripped and topped, the greater part of which were lodged. Moreover, the land on which this cane was grown was quite low, and the soil, a cold, clay loam, not well adapted for cane growing. Taking these facts in connection with the bad season, it must be looked upon as below the average yield.

DEVELOPMENT.

The development of the Early Amber cane raised on this farm may to some extent be seen from the following analyses, which have been made by me during the summer and fall:

August 10 :
    Cane sugar _____ 3.00
    Glucose _____ 4.50
August 20 :
    Cane sugar _____ 8.20
    Glucose _____ 5.10
September 6 :
    Cane sugar _____ 9.22
    Glucose _____ 4.20
September 14 :
    Cane sugar _____ 9.96
    Glucose _____ 3.45
September 17 :
    Cane sugar _____ 9.86
    Glucose _____ _____ 3.32
September 20 :
    Cane sugar _____:_____ 10.02
    Glucose _____ 3.23
September 22 :
    Cane sugar _____ 11.05
    Glucose _____ 2.60
September 29 :*
    Cane sugar _____ 8.59
    Glucose _____ 3.50
September 29 :*
    Cane sugar _____ 8.60
    Glucose _____ 3.50
September 29 :*
    Cane sugar _____ 8.61
    Glucose _____ 3.44
October 3 :
    Cane sugar _____ 12.67
    Glucose _____ 2.43

*This cane was lodged by storm.

From these we see that the cane sugar gradually and rapidly increased, while the glucose slowly decreased, from the time of flowering to the maturity of the seed. During the latter part of September, most of the cane was lodged by a very violent wind and rain storm. The juice from the stalks that were lodged was charged with a red coloring matter, the inside of the entire stalk being in many cases of a bright red color. Several of the stalks contained but a small portion of red coloring matter, but instead had a peculiar yellow and watery appearance, and quite a disagreeable taste. The juices from these contained on an average only 8 per cent. sugar, and 4.8 per cent. glucose.

EFFECT OF LEAVING CANE CUT IN THE FIELD.

A number of stalks still in good condition, the juice of which contained 9.50 cane sugar and 3.25 glucose, were cut and left in the field ten days, during almost constant rain. At the end of the ten days the juice contained 5.98 cane sugar and 6.15 glucose. Some Early Orange cane was also cut September 20, when the juice contained 10.50 cane sugar and 4.95 glucose, and was left in the field till November 2, when the juice contained 13.80 glucose, while not a trace of cane sugar was present. These experiments show conclusively that if cane is cut or injured and left exposed to rain, the destruction of cane sugar goes on very rapidly, being in time entirely changed into glucose. The rapidity of the change depends, of course, in great degree on the weather.

EFFECT OF LEAVING CANE CUT, UNDER SHELTER.

In order to ascertain the effect of leaving cane under cover, two tons of Early Amber cane were cut, the juice containing 10.02 per cent. of cane sugar and 3.23 per cent. of glucose. One-half was topped and stripped and both lots were placed on the floor of the barn. The change taking place may be seen from the following table:

	Cane sugar.	Glucose.
SEPTEMBER 20.		
The cane freshly cut	10.02	3.23
OCTOBER 4.		
After two weeks:		
(Stripped)	8.25	6.21
(Unstripped)	8.17	6.00
OCTOBER 19.		
After four weeks:		
(Stripped)	7.41	3.41
(Unstripped)	7.64	3.74
NOVEMBER 2.		
After 6 weeks:		
(Stripped)	8.26	3.74
DECEMBER 20.		
After 13 weeks:		
(Stripped)	8.45	6.80

To judge by the table the cane changes very slowly, but in reality the loss of sugar is quite rapid. If no loss of sugar took place, the juice would of course become richer in sugar, on account of the evaporation of part of the water. In reality this is not the case. The cane sugar becomes gradually changed to glucose, which in turn is destroyed by fermentation. In this way the juice may become even richer in sugar, but the quantity of juice is greatly diminished. The juice becomes also very acid. The effect produced by shocking the cane in the field was tried, with very unsatisfactory results, the cane sugar being destroyed very rapidly.

EFFECT OF LEAVING CANE STRIPPED IN THE FIELD.

One part of a patch of Minnesota Early Amber cane was stripped of leaves and left standing in the field from September 15 to September 22. It was then cut, and the juice, together with some that had not been stripped, was analyzed, with the following result:

	Cane sugar.	Glucose.
Cane stripped for one week	11.05	3.25
Same cane not stripped	12.98	2.78

The diminution of sugar is undoubtedly due to the fact that the latent leaf buds found under each leaf begin to develop into new leaves. These new leaves are formed partly at the expense of the sugar in the cane.

DEFECATION.

The juice after it leaves the mill has a more or less green color, due to the presence of large quantities of chlorophyl and other vegetable substances, which must be removed. This process is known as defecation.   The defecator, or the vessel in which this operation is conducted, may be of wood..   Copper is perhaps the best material, but is much more expensive.   The vessel should be furnished with a steam coil, with sufficient capacity to heat the juice to the boiling point in a short time.   As soon as the juice is expressed it should be removed to the defecator, where it should be heated at once to about 175° F., or just about hot enough to enable a man to hold his hand in the juice without being scalded.   Milk of lime, freed from all coarse particles by straining, should then be added until a slip of red litmus paper becomes changed to a faint purple when dipped into the juice.   The lime should be added in small portions, the juice being vigorously stirred with a paddle after each addition.   When the right quantity has been added, the juice must be heated as quickly as possible.   A thick green scum will soon come to the surface.   When the boiling point is reached—which is shown by the swelling and breaking up of the scum—the heat should be stopped and the juice left quiet for about five minutes.   The scum will then be quite hard, and may be easily removed from the surface of the clear liquid.   Much will depend on a good defecation.   If the defecation has been properly conducted, the liquid will be clear, free from particles, and of a pale yellow color.   If the scum is of a light color and thin, while the liquid below is opaque and has a greenish color, it shows that too little lime has been added; while if the juice is very dark, too much lime has been used.   Much nicety of judgment is required to make a good defecation, which can only be obtained by experience.

USE OF SULPHUROUS ACID.

The clear juice from the defecator is now tolerably pure, most of the impurities having been eliminated.   It contains, however, considerable lime, which if allowed to remain will give us a dark sirup, and if present in sufficient quantities will impart a more or less bitter taste to the sirup.   To avoid this we must neutralize the lime, just as before we neutralized the acid.   For this purpose sulphurous acid is much used.   This acid may be added to the juice in the defecator after removing the scum, or it may be added to the juice in the evaporating pan.   A sufficient quantity should be added to give to the juice a distinct acid reaction, or until a slip of blue litmus paper, dipped into the juice, is reddened.   To accomplish the same result, many preparations have been sold to the farmers and other sirup manufacturers by agents and peddlers.   I would here advise every one to leave all such preparations alone.   Most of them are either harmful or good for nothing, while others are but modifications of the methods which I have described and for which the buyer pays an exorbitant price.   As long as I remain at the university inquiries as to any method will be answered.   Before closing this report we will describe methods by which sulphurous acid may be made at sirup works.

BOILING TO SIRUP.

The juice should be boiled down as rapidly as possible, the scum which comes to the surface being skimmed off.   If conducted entirely in an open train it should be concentrated till it boils at about 234° F., which corresponds to about 45° B.   If a vacuum pan is used the sirup should be transferred to it when it has a density of about 20° B. It should then be concentrated to about 44° B., at as low a temperature and as quickly as possible.   If the sirup is made too thick, the crystals of sugar will be small and difficult to separate; while if too thin the crystallization will take place very slowly.   After the sirup has been boiled down to the proper density it should be placed in a room where the temperature may be maintained at about 90° F. to crystallize.   The crystallization usually begins in a few hours, and in five or six days the sugar may be separated.   The sirup may be boiled down a second time, and a second crop of crystals equal to about one-half the quantity of the first may be obtained in a couple of weeks.   A good yield of sugar may be obtained if the following rules are strictly adhered to:

1. Do not cut the cane until the seed begins to harden.
2. Do not allow the cane to stand stripped in the field.
3. Work up the cane as soon as possible after being cut.
4. Defecate the juice as soon as possible after leaving the mill.
5. For defecation use milk of lime, freed from coarse particles by straining; add it gradually to the juice with vigorous stirring until a piece of red litmus paper is turned to a pale purple.
6. Heat the juice quickly to the boiling point, as shown by the swelling and breaking of the scum.

7. Remove the scum after allowing the juice to remain quiet for five minutes.

8. Draw off the clear juice through an aperture near the bottom of the defecator into the evaporating pan.

9. Add sulphurous acid to the clear juice until a piece of blue litmus paper is reddened.*

10. Evaporate down until it reaches a density of 45° B., or, if boiled in an open pan, to a boiling temperature of 234° F.

11. Place in a warm room to crystallize, and in about a week it will be ready to separate

RESULTS.

Below will be found a table containing the summary of the results obtained from two plots. Plot A was planted with seed obtained from Mr. Seth Kinney, of Morristown, Minnesota. Plot B was planted with seed from Mr. Charles Eustis, of Fort Atkinson, Wisconsin. Plot A was very much exposed and a great deal of the cane was lodged, while Plot B was more sheltered and the cane was in better condition:

	Field of Plot A.	Field of one acre at the same rate as Plot A.	Field of Plot B.	Field of one acre at the same rate as Plot B.
Area of plots in acres	1³⁄₅			
Total weight of cane	4,669	30,348	4,710	23,550
Total weight of juice in cane	3,875	25,187	3,909	19,545
Weight of juice expressed	2,680	17,420	2,732	13,560
Weight of juice left in bagasse	1,195	7,767	1,177	5,885
Per cent. of cane sugar in juice	9.24		10.53	
Per cent. of glucose in juice	3.53		2.68	
Total weight of cane sugar in cane	358	2,327	415	2,075
Weight of cane sugar in expressed juice	248	1,612	290	1,450
Weight of cane sugar in bagasse	110	715	125	625
Weight of sirup obtained	332	2,158	408	2,040
Weight of cane sugar separated	142	923	199½	997½
Weight of molasses	190	1,235	208½	1,042½
Bushels of seed		27½		32

A glance at the table will show at once the wastefulness of the present mode of extracting the juice. Out of 85 per cent. in the cane, only 60 per cent. was obtained, or nearly 30 per cent. of the sugar in the cane was left in the bagasse. This loss is undoubtedly smaller than that sustained in the majority of cases, as 60 per cent. of juice is larger than the average per cent. obtained by the small mills usually employed. The absurd theory, that if too much juice is expressed it will cause the whole to "sour," make a poor sirup, &c., is entirely false.

THE DIFFUSION PROCESS.

The diffusion process for extracting the sugar from both beets and cane is now employed in nearly all of the principal factories. The cane is cut into thin slices by rapidly revolving cutting machines, the sugar being extracted from these by the use of water. If the pieces of cane are placed in a vessel and a quantity of water equal to the quantity of juice in them be added, part of the sugar will at once pass through the cell walls into the surrounding water, while part of the water will enter the cells. This will continue until the liquids inside and outside of the cell walls are of the same density. If this water be drained off it will contain half the sugar. If, now, this same cane be treated with equal and successive portions of water, each portion, when drained off, will contain one-half of the sugar contained in the cane at the time when it was added. In other words, the cane will retain after each draining one-half, one-fourth, one-eighth, one-sixteenth, one-thirty-second, etc., of the sugar originally in the cane. In practice this process is carried on in such a way that the water is used over again on successive portions of cane until it becomes nearly as rich in sugar as the juice, only about 20 per cent. of water being added. An apparatus working on this principle has been invented in Europe, in which slices of cane or beets are made to pass upward through a cylinder, by

*This step may be omitted if no excess of lime has been added during defecation. It will have no effect on the quantity of sugar obtained, but will make a lighter colored molasses.

the aid of a mechanical feeder, while water passes in at the top of the cylinder, and in passing down becomes more and more charged with sugar, until it issues from below, carrying with it almost the whole of the sugar from the cane.

In this way it is claimed 94 per cent. of all the sugar in the cane is obtained, or 24 per cent. more than that obtained by an average good mill. This difference would constitute an immense profit in a large establishment. The juice is, moreover, perfectly clear, containing but small quantities of chlorophyl and other vegetable matter, which occur so abundantly in juice expressed by the mill. A better sirup and a larger yield of sugar is the result.

CANE FOR SIRUP MAKING.

For the making of sirup exclusively some experiments were made with the Early Amber, Early Orange, and Honduras. Three plots were planted, one with each variety, in close proximity to each other. They received the same amount of cultivation, and the comparative results are, we believe, as fair as they can possibly be made. The plots were each one-fifth of an acre; and for convenience sake, the results in the following table are calculated to one acre:

	Early amber.	Early orange.	Hondu- ras.
Weight of stripped stalks...	23,520	31,000	42,330
Weight of juice expressed..	13,660	17,966	24,433
Per cent. of juice expressed...	58.80	57.95	57.70
Degree Baumé of juice...	8.0	8.5	7.0
Per cent. of cane sugar in juice	10.63	10.50	7.00
Per cent. of glucose in juice..	2.68	4.95	4.20
Gallons of sirup obtained...	180	239	265

There was no marked difference in the quality of these different kinds of sirup, and it would certainly repay the cane growers to try the Honduras as a sirup-producing cane. One great obstacle, however, is that the seed would have to be imported from more southern localities every season, as the seed hardly reaches beyond the milk stage before frost may be expected.

METHODS FOR MAKING SIRUP.

Several different methods for making sirup were used. The lightest colored sirup will be produced when the juice is nearly boiled down and skimmed without defecation. The acids which in that case remain free in the sirup change large quantities of the cane sugar to glucose and impart the "sorghum taste" to the sirup. In order to make a sirup free from this taste the juice must be defecated. The defecation should be conducted in the same manner as that described under sugar making. If too much lime is added a dark sirup will be the result. If the lime is added very carefully, so as to make the juice very nearly neutral, an excellent sirup will be produced. The following rule for defecating juice for sirup works well: Fill the defecator three-fourths full with fresh juice; heat to about 160° F., and add milk of lime perfectly freed from coarse particles, until the juice becomes slightly alkaline. Fill the defecator with fresh juice, mix well and heat to boiling, skim and boil down to a sirup. The defecation may also be carried out as described under sugar making, a quantity of sulphurous acid being added to the defecated juice until it becomes slightly acid. If properly conducted this process will always make a good sirup. It is probably to be preferred to any other, as it is very easily performed. Not much care is requisite, as any small excess of sulphurous acid which has been added will escape with the steam during the boiling down of the juice. Sulphate of aluminum may be used instead of sulphurous acid with equally good results, but more care is necessary, since any excess that is added will remain in the sirup. The flavor of the sirup will depend to a very great extent on the quantity of lime used for defecation, and the quantity to be added must be ascertained by practice. If the maker finds that the sirup still retains some of the "sorghum taste" it is a proof that too little lime has been used, and a stronger defecation should be made. If, on the other hand, the sirup is very dark, too much lime has been added.

CONSUMPTION AND PRODUCTION.

According to the late Commissioner of Agriculture a total of 2,000,000,000 pounds of sugar was consumed in the United States during the year 1879. "Of this amount

1,74:,560,000, or more than 80 per cent., besides 38,395,575 gallons of molasses, were imported. The whole valued at $114,516,745." He says further: "To bring the vast amount of sugar imported into this country within more easy comprehension, we have only to imagine five vessels of nearly 500 tons each and loaded with sugar, arriving at our ports each day in the year." The question, therefore, can sugar be profitably manufactured from northern sugar cane, is one of immense importance to this country. That there is much prejudice to be overcome is evident. There are men to whom the bare idea seems ridiculous. In the face of these difficulties, however, we venture to state that if skillfully conducted, the manufacture of sugar from this cane will certainly pay. Assuming the sugar to be worth 8 cents per pound, and the molasses 30 cents per gallon, we have the value of the produce per acre as follows:

Yield at the rate of plot A:

923 pounds of sugar at 8 cents	$73 84
103 gallons of sirup at 30 cents	30 90
Total	104 74

Yield at the rate of plot B:

997½ pounds of sugar at 8 cents	$79 80
87 gallons of sirup at 30 cents	26 10
Total	105 90

The seed has a composition about the same as corn, and will undoubtedly constitute a good food for farm animals. The utilization of the by-products will constitute another source of income. The first scums, being very rich in nitrogen and mineral salts, will make an excellent fertilizer, and from the last scums, being rich in sugar, a good vinegar may be manufactured. Taking also into consideration that my experiments were conducted on a small and consequently a wasteful scale, my results are undoubtedly too low. If the capital is sufficient to produce both refined sugar and sirup, the value of the products will be increased by at least one-third.

COST OF PRODUCTION.

The cost of production is of course the main consideration, and although I cannot as yet give any definite figures, I am confident that after paying all costs a good profit may be realized. The best plan for conducting this industry will be to have large central factories. During the working season these factories can work up a large quantity of cane grown in their vicinity, and during the remainder of the year the crude produce from smaller establishments may be worked up and refined.

SUCRATE OF LIME PROCESS.

The sucrate of lime process now in full operation in Europe seems to be eminently fitted for carrying out this plan. A very brief outline of the process will perhaps not be out of place here. Sucrate of lime is a solid, containing when dry about 70 per cent. of sugar, and having the appearance of sand. It is insoluble in cold water, but soluble in hot water, and also in solutions of sugar, not too concentrated. It is entirely unfermentable, and will not become mouldy or undergo decomposition, if kept for an indefinite length of time. It is therefore an excellent material for shipping and storing. Sucrate of lime may be manufactured on the farm with a comparatively small outlay. The juice is defecated as usual, and boiled down from 30° to 32° B. The sirup is then cooled and transferred to the sucration vessel. This vessel is usually made of galvanized sheet iron. In the center is a vertical shaft, carrying paddles. A certain quantity of pure and finely pulverized lime is then added, which becomes thoroughly mixed with the sirup by the motion of the paddles. The lime and sugar quickly combine, forming the sucrate of lime, which, when washed with cold water and dried, is ready for shipment to the refinery, where the sugar is separated from the lime and refined. This is, very briefly told, the process which we believe can be successfully applied to the manufacture of sugar from the sorghum cane. We trust that by another year, if these experiments are allowed to continue, some practical results in connection with this and the diffusion process may be brought out. It would have been very desirable to have made some experiments with these processes during the past season, but our time was entirely taken up by the work which has been done. Moreover, the limited amount of means at hand would not warrant the construction of the special machinery necessary for conducting these processes.

PRODUCTION OF SULPHUROUS ACID.

Considerable quantities of sulphurous acid are needed in making sirup, and much expense may be saved by making it at the factory. When sulphur is burnt in the air, each part of sulphur unites with two parts of oxygen from the atmosphere, forming a gas called sulphur dioxide. This gas is readily soluble in water.

When water has a temperature of 50° F., it will absorb 50 volumes, or one gallon of water will absorb 50 gallons of the gas. As the temperature of the water rises, it becomes less capable of absorbing the gas, so that at 70° F. it will absorb only 34 volumes. The solution of this gas in water constitutes sulphurous acid. Hence to prepare it, all that is necessary is to cause the fumes of burning sulphur to come into contact with water.

The easiest way for persons using steam-power to make this acid is to draw the fumes of burning sulphur from the furnace by a common gas pump and force them through a pipe reaching to the bottom of a barrel filled with water. The bubbles of gas escaping through the lower end of the pipe will be absorbed by the water in ascending. It is best to bend the pipe so that its lower end may lie along the bottom of the barrel. The open end should be closed, and the part lying on the bottom should be pierced with small holes so as to make a large number of small bubbles, instead of a few large ones, the gas being absorbed in this manner more rapidly. In this way a barrel of sulphurous acid may be made at a cost of from 75 cents to 80 cents. Any further information may be obtained on this subject by writing.

Below will be found the analysis of several bundles of cane, which I received from different parts of the State. Many bundles arrived without any labels, having lost them during transportation. Such samples were not analyzed, as it was impossible to tell whence they had been sent. If parties who have sent cane are not represented in the following table, it is because I have received no information in regard to the cane sent, or else the cane has been without labels, making it impossible for me to tell where it belonged.

ANALYSES.

Name of sender.	Name of grower.	Locality.	Character of soil.	Time of planting.	Date of cutting.	Number of stalks.	Weight of stalks.	Per cent. of juice expressed.	Per cent. of cane sugar in juice.	Per cent. of glucose in juice.	Time of analysis.	Variety.
A. C. Kent		Janesville	Sandy loam	May 15	Oct. 5	12	18	48.0	7.29	3.94	Oct. 8	Early Orange.
A. C. Kent		Janesville	Sandy loam	May 20	Oct. 5	12	12½	60.0	11.14	3.40	Oct. 8	Early Amber.
Wm. Toole	Alex. Toole	North Freedom	Clayey	May 15		12	8	53.1	6.50	5.45	Oct. 10	Early variety.
Wm. Toole	John Leifer	North Freedom	Clayey	June 23	Oct.	13	10½	58.8	9.92	2.96	Oct. 10	Early Amber.
Wm. Toole	Henry Rithe	North Freedom	Clay loam	May 26	Sept.	12	10½	52.0	6.25	5.12	Oct. 10	Early Amber.
Caspar Eberdt	C. Eberdt	Tomah	Sandy loam			12	12½	55.5	12.83	2.04	Oct. 10	Early Amber.
Caspar Eberdt	Griggs	Tomah				5	6½	55.5	10.94	3.59	Oct. 11	Early Amber.
Caspar Eberdt	Lene Sweet	Tomah	Heavy clay loam	May 15		12	10½	59.0	4.88	5.23	Oct. 11	Early Amber.
Caspar Eberdt	Forrest	Tomah				12	9½	56.0	8.98	3.45	Oct. 11	Early Amber.
Caspar Eberdt	Faulkner	Tomah	Sandy loam	May 20		12	12½	50.0	7.01	4.50	Oct. 11	Early Amber.
Caspar Eberdt	Prescott	Tomah				5	5½	61.5	6.77	5.78	Oct. 11	Early Amber.
Caspar Eberdt	Stevens	Tomah				11	13	52.0	7.08	5.36	Oct. 12	Early Amber.
L. B. Green	L. B. Green	Hebron	Light loam	May 15	Oct. 10	12	12	51.1	13.11	3.22	Oct. 12	Early Amber.
L. B. Green	L. B. Green	Hebron	Light loam	May 12	Oct. 10	12	11½	55.4	13.63	2.65	Oct. 12	Early Amber.
B. C. Dinsdale	J. Monteith	Preston	Black loam	May 26		12	15½	54.3	10.87	2.59	Oct. 12	Early Amber.
B. C. Dinsdale	Geo. Whitish	Preston	Clay		Oct. 6	12	8	58.3	6.00	8.43	Oct. 12	Early Amber.
B. C. Dinsdale	B. C. Dinsdale	Preston	Prairie loam	June 1	Oct.	12	12	58.1	6.23	7.56	Oct. 14	Early Amber.
E. W. Sanders	E. W. Sanders	Oshkosh	Sandy loam	May 5	Oct. 12	12	7½	60.7	6.21	3.98	Oct. 14	Early Amber.
J. W. Baily	J. W. Baily	Ripon	Clay land	June 1	Oct. 12	12	14	54.3	7.61	4.50	Oct. 14	Early Amber.
J. W. Baily	J. W. Baily	Ripon	Clay land	May 10	Sept. 24	12	8½	63.6	6.59	6.02	Oct. 14	Early Amber.
J. M. Nash	J. Daley	Hudson	Sandy loam	May 16	Sept. 28	12	11	71.7	10.31	4.70	Oct. 14	Early Amber.
J. M. Nash	E. L. Nash	Hudson	Black loam	May 12	Sept. 26	13	11½	66.6	9.07	5.36	Oct. 14	Early Orange.
J. M. Nash	J. M. Nash	Hudson	Very sandy	May 25	Oct.	12	13	57.7	4.50	9.50	Oct. 14	Stewart's Hybrid.
R. L. Clason	R. L. Clason	Beaver Dam	Clay	May 12	Oct. 8	6	6½	51.8	8.70	8.20	Oct. 21	Golden Imphee.
R. L. Clason	C. Driscll	Beaver Dam	Wash of barnyard	May 24	Oct.	9	4½	58.8	8.78	5.60	Oct. 21	Early Amber.
R. L. Clason	H. Lowell	Beaver Dam	Clay	May 15	Oct. 2	9	5	55.5	5.62	5.62	Oct. 21	Kansas Orange.
A. J. Decker	A. J. Decker	Fond du Lac	Clay soil	May 20	Sept. 10	12	14½	51.7	6.73	6.90	Nov. 2	Early Amber.
A. J. Decker	A. J. Decker	Fond du Lac	Black soil	May 20	Sept. 10	12	11½	58.7	6.43	5.52	Nov. 2	Early Amber.
A. J. Decker						12	15½	49.0	7.15	6.65	Nov. 2	Early Amber.

It is hardly possible to draw any definite conclusions from the above analyses, as many samples were not received for several weeks after being cut. It will be seen, however, that nearly all those samples which were analyzed within but a few days after being cut contain a large proportion of cane sugar, while those which were analyzed after a longer period of time show a high content of glucose and a low proportion of cane sugar. This corroborates my statement in the first part of this report, and shows the necessity of working up the cane directly from the field in order to get the best results.

It will also be seen that all the samples conspicuous for their high content of cane sugar are raised on a light soil, usually sandy loam, while those raised on heavy clay land contain large proportions of glucose. It therefore appears that in order to obtain a maximum content of cane sugar the cane should be grown on a light soil. For making sirup alone, the cane raised on clayey land will do about as well, as the high content of glucose will not materially affect the quality of the sirup.

VARIETIES OF CANE GROWN FOR EXPERIMENTS.

By Prof. W. A. HENRY.

Amber cane was grown from seed obtained from Charles Eustis, Fort Atkinson, Wis., and Seth Kinney, Morristown, Minn. From Mr. Kinney were also several packages of seed of Early Amber grown under different conditions. From J. A. Hedges, Saint Louis, Mo., Kansas Orange, Hedges' Early Orange, Early Orange and Honduras. Also Early Orange from Illinois Industrial University, Champaign, Ill., through Prof. M. A. Scoville.

All these varieties and sub-varieties showed peculiarities worthy of attention, but it is useless to report from one season only.

The experiments this year all centered about the question of how much sugar and sirup could be obtained from the cane, and in this Mr. Swenson's attention was so absorbed that the relative merits of each of the varieties could not be investigated.

A thick or thin stand of cane evidently makes a great difference in the quality of the juice, and a fair test of varieties can only be made when each has been planted in various ways as to width of rows, distance apart of hills and number of stalks in the hill.

There is no doubt but that varieties vary in value, and it is important that the peculiarities of each be known, yet it is a more difficult task to find this out than with most farm crops. If the experiments are continued next season, this will be one of the problems to work upon.

AMOUNT OF SIRUP PRODUCED PER ACRE.

As might be expected, the reports show a wide variation in the amount of sirup which is obtained from an acre of ground. Not only does the difference follow from variations in quantity and quality of cane produced, but also from varying densities to which the sirup is reduced. Some manufacturers make a much thicker sirup than others. The reported yields, therefore, show only in an imperfect way what can be obtained; still they are of value, I think, to those looking up the subject.

S. Hanson, of Whitewater, one of the oldest and most experienced growers in the State, reports 18 gallons from 10 rods of ground and 200 gallons per acre from larger pieces.

Joseph H. Osborn, Oshkosh, reports the highest yield, 226 gallons, with an average of 150.

N. D. Comstock, Arcadia, Trempealeau County, estimates the average at 125 gallons, Maxon and Almony, Milton Junction, Rock County, estimate the average at 150 gallons.

J. H. Rhodes, Sextonville, Richland County, raised on 1 acre 170 gallons.

O. S. Powell, of River Falls, Pierce County, estimates the average crop at 100 gallons.

H. T. Webster, Keene, Portage County, obtained 40 gallons from 28 rods of ground.

J. D. Sherwood, Dartford, Green Lake County, reports one third of an acre yielding 12,588 pounds of stalks, from which 79.14 gallons of sirup were made.

A. J. Decker, Fond du Lac, considers 125 gallons the average.

Mr. S. Nason, of Nasonville, Wood County, where cane was grown this season for the first time, reports 800 gallons from 4 acres.

Evan Erickson, Stevenstown, La Crosse County, obtained 1,050 gallons from 5 acres.

The average yield of sirup on good ground in a favorable season may be set down at about 160 gallons. With such culture as is usually given to it the yield will be about 100. It may be set down as a fact that wherever it has been planted in the State it has succeeded, no matter how poor the soil was. It promises to be one of the very best crops for our sandy lands, for though the yield per acre will not be large, the sirup will be of fine quality. Land on the experimental farm which produced 50 bushels of corn per acre this year gave 200 gallons of thick sirup.

TONS OF CANE PRODUCED PER ACRE.

This season several of the large manufacturers have purchased cane by the ton, the price paid usually being $2.50 per ton for stripped and topped cane delivered at the mill. This makes it important to ascertain the number of tons produced per acre.

I take the following yields from the same source as before:

N. D. Comstock, Arcadia, Trempealeau County, grew fifteen tons, yield 171 gallons, on one acre.

George Grant, Janesville, reports one instance of 11 tons, grown on an acre, producing 14 gallons of sirup per ton, each gallon weighing 11¼ pounds. A. C. Kent, Janesville, puts the average product for the year at 10 tons per acre. The average may be safely stated at from 10 to 12 tons per acre, according to the soil and season, I think. Should the industry grow in importance, purchasing cane by weight from the grower will become a very common practice, and if we may judge from the difficulties arising between beet growers and sugar manufacturers in France, it is easy to see that no small amount of trouble will occur with us.

To purchase cane simply by weight, without regard to its character, will be alike ruinous to manufacturer and grower. In some way the quality of the juice must be considered. For sirup making, a densimeter, as the Baumé scale, will do fairly well in helping determine the true value of cane. In the standard ton of cane the stalks should be straight, with leaves and top removed, all small canes and suckers being left out. The juice should have a certain density, as shown by the Baumé scale.

The price for such cane could be agreed upon by growers and manufacturers before planting time. At the same time the price of cane which falls below this standard or rises above it can also be arranged.

Those who are contemplating this business on a large scale cannot turn their attention to this part of the industry any too soon, for our farmers are too independent of any one crop to attempt raising Amber cane for sirup boilers who are so careless that they will not pay for what they get according to its true value. Great care must be exercised to make the business a profitable one for careful growers. By purchasing according to a standard, the grower who plants upon sandy land, for instance, and produces a very high grade cane, may find the small number of tons from an acre returning a good profit, while a stated price per ton, without regard to quality, would drive him from the business.

It may be interesting to note in this connection that in France the price is $4 for a ton (2,200 pounds) of beets, the juice of which has a density of 5.5° (1.035) and that for variation above or below this standard special contracts are usually made, though in general where the system has been adopted 80 cents is added to this price for each additional degree above the standard, and subtracted for each degree below.

CANE SEED FOR FEED.

For several reasons the value of cane seed for feed has received little attention. Its importance has not yet impressed itself upon cane growers. As will be seen from Mr. Swenson's report, from one-fifth of an acre of ground, 6¼ bushels of seed, weighing 53 pounds per bushel, were obtained, or at the rate of 32 bushels per acre.

The average yield of oats in the vicinity of Madison this season was about 35 bushels.

J. M. Edwards, Oak Hill, Jefferson County, reports 230 bushels of seed, weighing 58 pounds per bushel, from 9 acres.

I do not think the feeding value of this seed can fall below that of oats, and possibly it is nearly equal to corn. Experiments will be tried upon the farm this winter to learn its value by practical tests.

There is no difficulty in saving the seed, as the heads can lie upon the ground a long time unless there is an excessive amount of rain. The heads can be drawn and spread on the barn floor, or, what would be better, arranged on racks in a shed like broom corn. Some bind the heads in bundles and stand them on end in the field like bundles of wheat, to dry.

According to one test the weight of the green leaves as stripped from the cane is nearly one-fourth as much as the weight of the stripped cane. From this I estimate that an acre producing, for instance, 12 tons of stripped cane will yield 3 tons of green leaves, which will afford somewhere between half a ton and a ton of dried leaves per acre.

All who have fed these dried leaves speak of them as equal to hay in value; they are not difficult to dry or care for, but owing to the season of the year and the great press of work at that time they are apt to be neglected. The unusual rainy fall made it impossible to save the leaves from our cane for experimental feeding, as had been intended.

As a summary of the reports sent in by 180 manufacturers, I would state that the season, upon the whole, cannot be called a favorable one. Probably owing to the intense cold that came on in November, 1880, the vitality of cane seed was so injured that when planted last spring it failed in many instances to grow. This cut down the acreage very considerably in many localities. The fall frosts were long delayed, and in this regard the season was peculiarly favorable. The almost daily rains during the whole fall made stripping very disagreeable and the roads almost impassable, so that the cane could not be drawn far, and much of it spoiled in the fields. Again heavy autumn winds laid the cane flat and tangled it, making the expense of stripping and cutting fully double what it should have been.

Mr. Swenson's analyses show that the cane sugar is mostly changed to glucose when the cane is blown down, though the loss is not so manifest when sirup alone is made. Had sugar been the object with our manufacturers this season it would have been a very unfavorable one.

This year has seen the introduction of steam into quite a number of factories, by which means sirup can be made much cheaper than by direct heat. With such facilities defecation is easily practiced, and sirup of superior quality made. I consider the success attained by these steam boiling works as the most marked event of the season. Previous to this year no one had but a few hundred dollars invested in the business. There seemed to be no chance for capital to take hold of it as long as direct heat was used, but with the introduction of steam apparatus capital can be invested with profit to the owner and advancement to the business. With so many large manufacturers in the field Amber sirup must go into the market in considerable quantities, and this, with the high quality of the goods, will soon command public attention. It is the introduction of these large factories that we must expect and encourage, if this is to become one of the great industries of the State.

One of the plainest lessons of the season is the importance of growing cane close to where it is worked up. A wagon load of the stripped stalks at the crusher is not worth over $5.

It at once becomes evident that such weighty material cannot be drawn long distances with any profit, and that the sirup works must be located near the fields where the cane is grown. Cane to be profitable should not be grown over two miles from the works, unless the roads are excellent, when possibly three may be set as the limit. Those who are locating mills should aim to settle at points where the cane fields can be about them on all sides. Fuel need not be considered, for the bagasse is sufficient when properly managed to supply all the heat needed. The transportation of the sirup requires that the works be near a railroad station.

Another fact of the utmost importance has been made plain this season, that is, defecation of the juice by some method is essential. The prejudice against the sirup because of its acid or "sorghum taste" keeps the market price down below what it should be, and then buyers will only take it at a low price or not at all. If they must pay sirup prices, they prefer New Orleans.

Even the sirup shipped is not sold to the consumer direct, but is first mixed with glucose to remove the strong taste, or rather to favor the glucose.

The only way to overcome this prejudice is to make a sirup with the sorghum taste left out. The experiments on the farm and by others show this to be possible, and that the methods are, upon the whole, very simple. I am aware that quite a prejudice exists among boilers against any clarification of the juice. Some even argue that people refuse to purchase Amber cane sirup, not because of its sorghum taste, but because it is a home product. They forget that maple sirup, a home product, brings three times the price of the New Orleans.

Our boilers here exerted every effort toward making a light, colored sirup, and because lime darkens it they are afraid to use lime. If every boiler would use lime cautiously next season, letting color be considered after flavor, there would be more real advancement in the industry than ten years of present methods of attempted improvement will bring. At present less than 10 per cent. of the boilers use lime or practice defecation of any kind.

CAN THE FARMER MAKE HIS OWN SUGAR?

This is a question naturally asked by many who have not studied the problem to any extent.

Most certainly not, if profit is to be considered. A farmer might have a mill and make his own patent process flour, but it would not pay him. His business is rather to grow the wheat, while skilled men attend to the milling.

While first class Amber sirup can be made by proper means with a small investment and a fair amount of skill, sugar making must be left to skilled men under the direction of a chemist or expert. Such experts must be trained to work with northern cane, and not brought from southern localities where the conditions are very different. Such persons though experts at home would only be students, for a time at least, at the north. In order to manufacture sugar there must be quite a large investment of capital in machinery; to manage this there must be skilled men, and over all must be a man who by chemical tests reads the varying conditions of the juice as it runs from the crusher from day to day, and whose work is law with all other employés. Until there are such experts capital should be most cautious. Fine sugar works with costly machinery will not alone bring sugar, as the many past failures show. It would be far better for all concerned to wait ten years before another step is taken in this promising industry than to have it blighted in the start by failures. With capital carefully invested in proper machinery, the works located in the midst of cane fields, and run by good workmen and a skilled chemist, there is no doubt but money can be made as rapidly as in any manufacturing business. When success comes, the farmer will sell his cane at the sugar works as he does his wheat at the mill, but he will not be a sugar boiler and farmer combined.

EXPERIMENT WITH FERTILIZERS.

In order to ascertain the value of fertilizers in the production of sirup, an experiment was planned a year ago, in which the co-operation of our Wisconsin farmers was solicited. Over forty farmers agreed to carry out a simple experiment as I directed. The following are the directions which were sent to each in April last:

DIRECTIONS FOR THE EXPERIMENT.

Select in the field where cane is to be planted three plots of ground, each containing not less than ten square rods and lying side by side. The ground should be as uniform as possible in its composition and fertility. Do not select soil where one end of the plot is sand and the other loam or clay. No matter which it is, but have it all one character. Have the plots, if possible, long and narrow, say one rod by ten, or two by twenty, etc. The plots should lie side by side and should not be separated from one another or the rest of the field. One plot, No. 1, plow in well rotted stable manure at the rate of sixteen large loads per acre—one load for every ten rods. Plot No. 2, which is to be the middle plot, has no manure of any kind upon it. When the cane on plot No. 3 is three or four inches high apply plaster to the hills or rows to the amount of 160 pounds per acre, or 10 pounds for every ten rods. The cane is to be planted and cultivated in the same manner as the rest of the field. If possible, weigh the cane of each plot separately when ready for the mill. Boil the juice to a sirup weighing 11½ pounds per gallon, and determine accurately the yield of each plot. Save a sample of sirup from each plot for comparison.

Report to the department upon the following points:

1. Amount of ground in each plot.
2. Character of soil—clay, loam, sand, etc.
3. Is soil naturally rich or poor?
4. Number of years the field has been in cultivation.
5. Crops grown on field previous year.
6. Whether or not the field was manured the previous year.
7. Method of planting cane—in drills or hills.
8. Time of planting.
9. Time of ripening.
10. When manufactured.
11. Yield of sirup from each plot.
12. Character of sirup from each plot as to color, clearness, and flavor.

But one of all who agreed to undertake the experiment carried it through successfully. Mr. S. B. Chatfield, of Adams, Walworth County, makes the following report:

ADAMS, *January* 2, 1882.

DEAR SIR: I have been so very busy that I have neglected to send samples until to-day. I express them as you requested. I will answer those questions to the best of my ability:

No. 1. One rod wide, 10 rods long.
No. 2. Black sandy loam.
No. 3. Naturally rich.
No. 4. Under cultivation 33 years.
No. 5. Sugar-cane.

No. 6. Not manured the previous year.
No. 7. In drills.
No. 8. Planted 19th of May.
No. 9. Ripe from 12th to 15th of September.
No. 10. Manufactured September 28.
No. 11. No. 1, 17 gallons; No. 2, 10 gallons; No. 3, 14 gallons.
No. 12. The three samples must speak for themselves.

Mr. W. A. HENRY.

The samples were indeed interesting. That from unmanured soil was light-colored, and sugar crystals in considerable numbers and of fair size formed in it. The sirup from the manured plot was the darkest. Other qualities, marked in their way, I am very sorry I cannot report on, as Mr. Chatfield's samples were put on exhibition at the State cane-growers' convention, and two of the bottles were carried off by some visitor.

It is unfortunate that more had not been as persistent as Mr. Chatfield, for untold good would flow from united work in this way.

If there are any of our farmers who are willing to try such an experiment again, I shall be pleased to have their names and will forward directions in due time.

The importance of united work will appear plain to all who have grown cane to any extent.

LIST OF SIRUP MANUFACTURERS IN WISCONSIN.

The following is a list of all manufacturers whose names I have been able to obtain, together with address and amount of sirup made by each during the fall of 1881.

For convenience of reference, they are arranged alphabetically by counties:

Name.	Post-office.	County.	Gallons sirup made in 1881.
George Cochran	Gilmanton	Buffalo	1,350
Edwin Blood	Stockbridge	Calumet	7,700
John B. Sweet	Chilton	Calumet	900
C. C. Carr	Poynette	Columbia	1,700
L. K. Goodall	Lodi	Columbia	320
Charles W. Peters	Columbus	Columbia	863
Cyrus Root	Otsego	Columbia	1,000
L. S. Wright	Fall River	Columbia	2,000
I. B. Hayden	Freeman	Crawford	1,000
Samuel A. Clark	Prairie du Chien	Crawford	1,000
W. J. Lankford	Ferryville	Crawford	800
C. R. Rounds	Mount Sterling	Crawford	1,000
A. H. Anderson	Black Earth	Dane	1,200
J. H. Greening	Mazomanie	Dane	2,002
Henry Linley	Mazomanie	Dane	1,200
W. M. Sprague	Lake View	Dane	300
B. F. Williamson	Madison	Dane	1,350
R. L. Clason	Beaver Dam	Dodge	2,200
C. J. Davis	Beaver Dam	Dodge	853
Charles Link	Danville	Dodge	1,800
Joseph Philips	Randolf	Dodge	1,400
W. H. Clyde	Rock Falls	Dunn	2,000
W. H. Doane	Fall City	Dunn	800
H. J. Myers	Elk Mound	Dunn	1,878
F. M. Steves	Louisville	Dunn	4,100
W. W. Waterbury	Augusta	Eau Claire	2,600
George W. Jones	Fairchild	Eau Claire	800
J. W. Bailey	Ripon	Fond du Lac	3,000
A. J. Decker	Oakfield	Fond du Lac	10,000
C. J. Gordon	Oakfield	Fond du Lac	5,800
George Jenkinson	Brandon	Fond du Lac	2,500
M. M. Alexander	Montfort	Grant	550
C. D. Barnes	Brodtville	Grant	1,530
Francis A. Markert	Lancaster	Grant	125
Lewis Glass	Wyalusing	Grant	105
C. S. Ruddock	Berlin	Green Lake	1,500
G. W. Sheldon	Markesan	Green Lake	3,700
J. D. Sherwood	Dartford	Green Lake	4,877
Aug. Ziemer	Berlin	Green Lake	1,830
Peter Crook	Dodgeville	Iowa	900
J. P. Beard	Elroy	Juneau	650
F. W. Board	Elroy	Juneau	800
E. G. Dodge	Mauston	Juneau	2,400
Riley Moulton	New Lisbon	Juneau	1,500

Name.	Post-office.	County.	Gallons sirup made in 1881.
D. Travis	Wonewoc	Juneau	662
A. L. White	Mauston	Juneau	2,800
Wm. Goudre	Milford	Jefferson	900
P. W. & C. S. Cartwright	Rome	Jefferson	2,500
F. E. Chartier	Rome	Jefferson	5,000
E. Colwell	Farmington	Jefferson	2,500
J. M. Edwards	Oak Hill	Jefferson	4,200
John Moore	Rome	Jefferson	4,200
R. S. Pearsall	Waterloo	Jefferson	2,000
L. B. Green	Hebron	Jefferson	1,600
Frank C. Lehman	Watertown	Jefferson	2,800
Williams & Colwell	Farmington	Jefferson	8,000
Williams & Dow	Palmyra	Jefferson	900
W. H. Peardon	Palmyra	Jefferson	2,000
William Jaudre	Palmyra	Jefferson	1,000
Smith Hoyt	Milford	Jefferson	2,300
Geo. E. Allen	Milford	Jefferson	100
H. C. Davis	Irving	Jackson	1,700
B. C. Henry	Pine Hill	Jackson	1,959
L. W. Thayer	Kenosha	Kenosha	1,564
James F. Petrie	Kenosha	Kenosha	400
Evan Erickson	Stevenstown	La Crosse	1,800
Nels Hanson	Rockland	La Crosse	600
N. D. Loomis	West Salem	La Crosse	3,500
T. O. Masher	Bangor	La Crosse	000
Hollister Phillips	Mindora	La Crosse	600
Henry Rhode	Barre Mills	La Crosse	1,000
H. H. Slye	Mindora	La Crosse	2,025
Frank Pfaff	Burr Oak	La Crosse	1,165
Riley T. Scott	Yellowstone	La Fayette	1,465
Vincent Bruner	Blanchardville	La Fayette	550
Richard Graham	Jeddo	Marquette	1,876
G. A. Scott	Westfield	Marquette	2,700
T. Wells	Neshkora	Marquette	1,700
L. Baring	Oil City	Monroe	666
Casper Eberdt	Tomah	Monroe	1,800
W. G. West	Sparta	Monroe	1,675
M. Shidle	Sparta	Monroe	2,000
Samuel Thompson	Osceola Mills	Polk	3,209
W. H. Tilton	Osceola Mills	Polk	1,084
J. McLean	Saint Croix Falls	Polk	919
L. E. Buck	Sherman	Portage	1,000
W. M. Burroughs	Almond	Portage	2,500
Silas D. Clark	Plover	Portage	2,130
Nicholas Piper	Almond	Portage	2,576
Albert Taylor	Blaine	Portage	2,630
Reuben Thompson	Amherst	Portage	2,353
H. T. Webster	Keene	Portage	1,500
Alex. G. Coffin	Durand	Pepin	1,969
A. H. Cott	Jeddo	Pepin	1,700
D. W. Phelps	Durand	Pepin	900
S. L. Plummer	Arkansaw	Pepin	3,400
Hiram B. Stone	Durand,	Pepin	1,950
T. J. Atwater	Prescott	Pierce	2,000
O. S. Powell	River Falls	Pierce	10,500
L. L. Richardson	Clifton Mills	Pierce	2,000
Conrad Weghorn	Ellsworth	Pierce	2,100
Charles N. Soule	Rochester	Racine	400
Thos. McFarland	Waterford	Racine	1,200
Nims & Voorhees	Burlington	Racine	1,600
A. A. Cowey	Port Andrew	Richland	1,600
John J. Dillon	Basswood	Richland	1,000
R. W. Peters	Basswood	Richland	1,942
J. H. Rhodes	Sextonville	Richland	260
N. G. Sornum	Basswood	Richland	500
Thos. S. Palmer	Eagle Corners	Richland	400
Buob & Russell	Janesville	Rock	2,390
Conrad & Dibble	Evansville	Rock	2,000
George Grant	Janesville	Rock	2,100
A. C. Kent	Janesville	Rock	9,000
M. M. Tullar	Evansville	Rock	630
W. J. McIntyre	Whitewater	Rock	2,500
Maxon & Almony	Milton Junction	Rock	6,000
Bauernfeind & Alletzan	Glenbeulah	Sheboygan	4,000
M. J. Adams	Baraboo	Sauk	1,500
L. T. Allbe	North Freedom	Sauk	1,550
Isaac W. Carpenter	White Mound	Sauk	1,000
C. H. Dome	Baraboo	Sauk	1,140
G. F. Fuller	Baraboo	Sauk	1,100

Name.	Post-office.	County.	Gallons sirup made in 1881.
C. Henneberg	La Valle	Sauk	2,300
J. T. Huntington	Delton	Sauk	3,200
W. Jefry	Baraboo	Sauk	700
W. H. Koukel	Baraboo	Sauk	1,558
J. W. Shourds	Reedsburg	Sauk	1,100
C. R. Thayer	Baraboo	Sauk	1,410
R. F. Cole	Reedsburg	Sauk	1,460
J. B. Filbian	Hammond	Saint Croix	3,000
Foster & Nye	New Richmond	Saint Croix	3,500
F. W. Hitchings	N. Wis. Junction	Saint Croix	1,400
J. M. Nash	Hudson	Saint Croix	5,000
E. G. Partridge	Warren	Saint Croix	3,200
N. D. Comstock	Arcadia	Trempealeau	2,000
B. Dissmore	Whitehall	Trempealeau	2,050
A. F. Heusel	Arcadia	Trempealeau	450
A. H. Rogers	Osseo	Trempealeau	1,765
D. S. Watson	Whitehall	Trempealeau	2,800
H. H. Morgan	Red Mound	Vernon	3,600
L. F. Day	Retreat	Vernon	1,200
W. Frazier	Enterprise	Vernon	2,300
E. B. Hyde	Retreat	Vernon	900
M. K. Jefferies	Hillsboro	Vernon	225
C. Bloeman	Red Mound	Vernon	1,780
S. H. Helmer	Hartford	Washington	600
S. S. Nason	Nasonville	Wood	1,500
Henry Hull	Eureka	Winnebago	1,743
A. G. Lull	Oshkosh	Winnebago	800
Whitemarsh & Edwards	Oshkosh	Winnebago	2,000
Joseph H. Osborn	Oshkosh	Winnebago	3,500
W. M. Davies	Wild Rose	Waushara	1,150
Charles O. Dill	Oasis	Waushara	1,000
D. A. O. McGowan	Hamilton's Mills	Waushara	516
Wm. Scobie	Spring Lake	Waushara	1,600
H. C. Van Airsdale	Saxville	Waushara	1,000
W. M. Ware	Hancock	Waushara	777
Wilfred Lane	Wild Rose	Waushara	500
M. D. Morrison	Eagle	Waukesha	800
Romeo Sprague	Eagle	Waukesha	1,000
Edward P. Hinkley	Eagle	Waukesha	750
S. B. Chatfield	Adams	Walworth	1,606
T. M. Cook	Little Prairie	Walworth	1,150
S. Hanson	Whitewater	Walworth	2,000
Chas. E. Horton	Whitewater	Walworth	100
J. Patchin	Heart Prairie	Walworth	1,100
Pliny Potter	Little Prairie	Walworth	250
T. M. Shoudy	Geneva	Walworth	400
Ambrose Warner	Whitewater	Walworth	2,300
Richard Chambers	Weyauwega	Waupaca	550
John Clark	Waupaca	Waupaca	2,000
W. E. Clark	Bear Creek	Waupaca	900
R. J. Folks	Waupaca	Waupaca	2,300
E. G. Furlong	Rural	Waupaca	2,000
T. S. Neyward	Rural	Waupaca	1,800
Sumner Packard	Crystal Lake	Waupaca	1,000
Alvin Pope	Lind	Waupaca	3,000
J. Rode	Ogdensburg	Waupaca	400
Milton Stanley	Manawa	Waupaca	1,500
Total			340,610

The following names have been received since tabulating the above:

Names.	Post-office.	County.	Gallons sirup made in 1881.
Silas Hammond	Strong's Prairie	Adams	330
M. P. Hammond	Strong's Prairie	Adams	900
D. McDonald	Verona	Dane	800
Cyrus G. Patton	Augusta	Eau Claire	1,300
Gusta Yiss	Otto Creek	Eau Claire	1,400
John R. Roth	Platteville	Grant	1,000
Charles E. Bowerman	Patch Grove	Grant	718
James F. Brown	Mineral Point	Iowa	1,100
J. E. Arnold	Melrose	Jackson	800
R. Grant	Necedah	Juneau	618¾
L. F. Crandall	North Bend	Jackson	3,470
Wm. Caven	Mindora	La Crosse	2,700
Otto Amundson	Stevenstown	La Crosse	1,000
James Sykes	Stevenstown	La Crosse	1,100
John C. O'Bleness	Jeddo	Marquette	1,100
A. J. Cunningham	Woodstock	Richland	900
Travers & Snyder	Woodstock	Richland	1,500
Bennett & Mecum	Richland Center	Richland	800
Ole O. Lamb	Glasgow	Trempealeau	2,300
Alex. Sauer	Ettrick	Trempealeau	2,000
R. F. Gale	Reedsburg	Sauk	1,460
Charles Fuchs	Spring Green	Sauk	2,800
Jacob Isann	Spring Green	Sauk	500
Ole Kanteson	Spring Green	Sauk	500
William Stevenson	De Soto	Vernon	2,939
George C. Clark	Victory	Vernon	1,400
W. W. Minor	Retreat	Vernon	2,100
Joseph Morgan	Retreat	Vernon	1,800
S. M. Monaker	Liberty Pole	Vernon	2,000
Warren C. Bates	Retreat	Vernon	750
A. H. Bates	Retreat	Vernon	1,400
Lester N. Porter	Wautoma	Waushara	1,572

CORRESPONDENCE.

From among a large number of letters upon the subject, I select the following, which will, I am certain, be read with interest:

[From A. J. Decker, Esq , Fond du Lac, Wis.]

FOND DU LAC, WIS., *December* 17, 1881.

DEAR SIR: Another season has passed, and another harvest has been gathered with its lessons of success or failure. That should teach us in future years how to attain success and avert the chances of failure.

Though the past season has been the poorest in many years for growing Amber cane, and its manufacture into sirup and sugar, yet I think we have advanced very materially.

The late cold, wet spring greatly retarded planting, and fully one-third of the amount planted came up so poorly that it was plowed up and other crops planted. This was the case mostly with farmers who had little or no experience in raising cane, and mistook it for pigeon grass, or thought it looked too small to ever pay for the taking care of it, while farmers understanding it better cultivated it carefully and were paid with good crops. The fall has been very bad for the manufacture of sirup. The grinding season commenced about September 15, and by the 25th it commenced raining and rained almost every day for six weeks, until the country was flooded and roads impassable, some farmers feeding their cane to their cattle, a few of them storing it in their barns, hoping for better weather to haul it to the mill; and after I had finished the cane at the mill and had been shut down nearly a month, I started up to accommodate those farmers and to determine the amount and quality of sirup that could be made from cane kept so long after being cut, which was seven weeks. The result was a fine, light sirup, and about three-fourths of a full crop. Out of this lot was one-half acre from which I made 95½ gallons of sirup, for which the owner was offered 60 cents per gallon at the mill, which speaks well for its quality.

From the unfavorable season we have learned many valuable lessons which a favorable season would not have shown, and solving such difficult problems is taking a firm step in advance towards the time when this industry, with the aid of your Department, is to be an established source of business and wealth to the people of the State of Wisconsin.

One great drawback has been the lack of proper knowledge in the manipulation of the juice to obtain the best results, and people starting factories have been so anxious to get such information that they have been the easy prey of traveling sharks, claiming to be experts in the business, referring to some successful factory to which their name may be attached in some capacity, claiming by their skill and superior articles to have accomplished such results, and offering to sell a mill and outfit, for which they ask a fancy price, and will then give full instructions in their secret processes for 1 cent per gallon on each gallon of sirup made by them during the season. The work of your department will put a stop to this swindling business, and I hope the legislature will appropriate such amounts as may be requisite to fully develop the cane resources and place Wisconsin in a position to raise her own sugar and sirup, for which she has paid over $8,000,000 per annum. My factory has an easy capacity of 400 gallons sirup in twenty-four hours. I use steam for defecating and evaporating, and the Plantation Mill made by the Madison Manufacturing Company, and no other State can furnish a better one. I would be glad to have you visit my factory in grinding season if possible. Hoping for your complete success in developing the sugar resources of Wisconsin.

I am yours, truly,

A. J. DECKER.

Prof. W. A. HENRY, *Agricultural Department, University of Wisconsin.*

[From J. T. Huntington, esq., Delton, Wis.]

DEAR SIR: In reply to your request for something from me on the cane business, I submit the following:

The last two seasons have undoubtedly been unfavorable for the best results from Amber cane—the season of 1881 particularly as to yield in this vicinity. Notwithstanding that the season was very wet the yield of juice was generally small, but mostly of fine quality, my experience being that the juice of this year worked satisfactorily—much easier than that of last. The sirup from my works this year was, for a custom-mill where cane of all sorts is handled, very uniform in quality and color. We have in this vicinity all kinds of soil, and so far as I am able as yet to judge the very best results are obtained from cane grown on soil somewhat sandy, and if possible I would wish it to be on a clover sod. The finest flavored sirup and quickest to granulate of any made at my place are those from cane grown on a clover sod. Growers of cane, as a general thing, I think, do not do as they should to obtain the best results. Cane is too apt to be left to be the last thing planted and cultivated, and I have often had men tell me that they had only cultivated it once, and some not at all. Such cane cannot be satisfactory.

In my opinion, cane should be planted just as early as the climate will admit, covering just as light as possible, and cultivating as soon as the rows can be seen; and continue the cultivation until it is waist high, and then keeping the weeds out in August with a hoe.

It should be cut when a majority of the seed is ripe enough to grow, and if it cannot be worked at once, should be so placed that it can have *plenty of air*, and be covered from the rays of the sun or storms; so placed, it will keep well for some time. I have worked some that had been cut four weeks, and it was not at all soured—had, perhaps, lost a small portion of the juice. A matter of importance to manufacturers is a better market or better prices. The name generally applied is sufficient alone to make many refuse to purchase. At a time when ordinary New Orleans molasses is worth fifty to fifty-five cents in Chicago, at wholesale, forty cents is considered sufficient to pay for "sorghum," when the fact is that the "sorghum"'(when good as it ought to be) is the best goods to be had in the molasses line; and it is also a fact that large quantities of it (some not very good) are purchased in Chicago at very low prices, put into large tanks, and a little very *rank* New Orleans molasses added to give a New Orleans flavor, and then it is rebarreled and sold in the country as genuine New Orleans molasses. Probably those who will not buy "sorghum" direct of the maker often get it this way. There ought to be a manufacturers' association to work in their interests.

Yours truly,

J. T. HUNTINGTON,
Delton, Wis.

Prof. W. A. HENRY.

[A letter from Mr. William P. Phillips, of Lake Mills, shows that all do not look upon this question in the same light. Mr. Phillips writes as follows:]

LAKE MILLS, WIS., *December 12*, 1881.

DEAR SIR: Your circular of November 10, ultimo, relative to the Amber cane industry of Wisconsin, received. I am not in any manner interested in that branch of industry and know of no thrifty or practical farmer in this vicinity who is. Its production here is generally confined to a few of the smaller farms—usually those occupied by the poorer and most thriftless class of foreign-born immigrants—who are willing to use an inferior sirup of their own production, under the delusion that their time and labor in producing it is worth nothing. Only a few square rods are raised on each farm; and I apprehend if the labor in its production and manufacture was counted at its value in other established practical lines of agricultural business it would be found to cost many times the market value of much better sirup. In the present stage of development of the crystallizing process I am unable to appreciate the extraordinary efforts of the national and State departments of agriculture to foster its growth, or to obtain statistics in regard to it. It occurs to me that there are several things connected with the agricultural interests of this country in which the national and State departments—with their aided facilities—might do great service to the country.

We have established, partially developed, practical, and profitable industries that need the aid and benefit of the practical experiments of the departments and the protection of the Covernment.

Take as an instance the leading agricultural industry of our State—the dairy industry. Base, unwholesome, disgusting adulterations of dairy products are allowed to be manufactured and sold; our reputation and markets lost, or at least damaged at home and abroad. Millions are thereby lost to the farmers that a few unscrupulous persons, worse than counterfeiters, may defraud consumers out of a few thousands. Yet there has been no effectual law devised or passed, no effort worthy of the name been made to prevent or check the evil. The farmers, an unorganized class, are not capable of helping themselves. The State department of agriculture, as the only organized representative and guardian of the agricultural interests of the State, should repeatedly urge and secure the legislation required in this matter. The law on this subject passed last winter (chapter 40) accomplished nothing, as it was evidently intended it should accomplish nothing.

Again, the science of agriculture is yet comparatively undeveloped. True, it has made great advances in this country during the last half century, mainly by the knowledge gained by the experiments of private individuals. Like all sciences, money generally precedes experimental demonstration. To the private citizen experimental demonstration is often expensive or impracticable for the lack of facilities. The State department of agriculture should have some system of direct communication with the practical agriculturists of the State, by which inquiries might be solicited and answered, and the necessary experiments made at the expense of the State. An agricultural newspaper connected with the department might answer the purpose and be at least partially self-sustaining.

For instance, at the present time our stock and dairy interests require an immediate answer to the question of the economy and practicability of the preservation and use of ensilage as food for stock. We want no floating rumors picked here and there, but an authoritative answer based on the demonstration of reliable experiment.

Thus indefinitely questions daily present themselves to the practical farmer, and if you will inaugurate a system by which they may be satisfactorily answered by the department of agriculture you will greatly benefit the agricultural industries of the State.

I am, very respectfully,

WM. P. PHILLIPS.

rof W. A. HENRY, *University of Wisconsin, Madison, Wis.*

[From A. J. Russell, President Wisconsin State Cane Growers' Association.]

JANESVILLE, WIS., *December 19*, 1881.

DEAR SIR: In reply to your favor of the 8th, I would say that we have not purchased cane by the ton heretofore, as there was no reliable data to enable the manufacturer to determine the value of the different qualities of cane that was produced on different soils, and delivered at the mill in various conditions.

An imperfect knowledge and no well-developed system of determining the true value of the canes as delivered promiscuously from a large variety of soils has resulted in very serious losses to several large establishments who had adopted the method of purchasing cane delivered at the mill at a stipulated, and generally a uniform price per ton, or by the acre, irrespective of the purity of the juice contained in such canes.

There seems to be but one practical business method for a manufacturer to adopt for his own protection, and a greater satisfaction to the growers, and that is to purchase the cane by the ton. The manufacturer then has control of all the sirup and sugar, and is not brought into competition in the local or general market with his own patrons who grow the cane, many of whom have more than sufficient to supply their own and neighbors' wants, and desire to dispose of the balance they have on hand as soon as possible; and not being (as a general rule) familiar with the ruling prices of same class of goods in the wholesale and retail market, are imposed upon by dealers who are perfectly aware of the fact that the grower has not a sufficient amount to ship to jobbing points, and rather than hold it will sell it at a price to the local dealer generally below the actual market value, and that makes the price for manufacturers to the local trade as long as the grower's sirup holds out.

We have determined in the future to purchase our cane by the ton, delivered at the mill, and when so delivered will test the juice in the presence of the grower, and purchase it from him, same as grain and other farm products are purchased, according to quality. The actual value of the cane will be determined by the quality of the juice, and will be worth to the manufacturer from $1.50 to $4.50 per ton, and even $5 per ton for extra cane, and according to the state of the sirup and sugar market and the different degrees of purity of the juice, and the amount of sucrose contained in the raw juice at the time of delivery of the cane at the mill.

Our custom has been to charge the growers 25 cents per gallon, or one-half of the sirup.

Our works consist of a storage-room 20 by 40 feet, one story, shingle roof building, attached to our defecating, evaporating, and finishing building, which is 20 by 20, two stories high, and a shed attached for cane-mill, boiler, and engine.

Our machinery consists of boiler, engine, mill, juice-tank, juice-pump, defecators, evaporators, finishing-pan, cooler, and storage tanks.

The juice runs directly from the mill to the juice-tank, and is pumped up to the top floor into the defecator, and after the defecation is made it is discharged directly into the evaporator and rapidly reduced to a thin semi-sirup, and is then discharged into the finishing-pan and concentrated rapidly, if for sirup, to a commercial density and drawn off into the cooler, and almost immediately discharged into storage tanks sufficiently large to hold, each one of them, a little over a car-load.

When enough has been made for a car-load the barrels, three of them at a time, are rolled under faucets and filled. In that way it does not take us long to fill enough barrels for a car-load. We then ship generally to a wholesale market. Thus we have a continuous fall from the defecator to the barrels, without any rehandling of the sirups; and by cooling the sirup at once, after discharging into the cooler, it prevents the sirup from darkening by being sirup-scorched in running a succession of batches of hot sirup into a tank at a high temperature of heat so long that it darkens the sirup and lessens its value as a commercial article.

Our machinery is constructed and arranged to save labor and more perfectly clarify the juice and hasten the evaporation in the most rapid manner. Our defecators are so arranged and constructed that we do not have to skim the juice in them, and a simple attachment we have permits drawing the juice into the evaporator as clear as water. The knowledge of the fact gained by our own practical experience that the success of making a bright, glossy sugar and a light-colored, clear, transparent sirup, "without" the use of the expensive "char-filters," depended upon a perfect defecation, and a rapid concentration of the juice to the required density enabled us to build a style of evaporator that has produced the desired result, by enabling us to concentrate the juice rapidly, and at the same time liberate certain impurities that can be eliminated in no other manner known to us but by the application of heat; and when those impurities are separated and thrown to the surface they flow rapidly to the automatic skimmer and filter, where they are retained and forced over into the scum-trough in a comparatively dry condition, and the strained and filtered juice passing through the filter rapidly is returned immediately to the evaporator, again clear and transparent. In this manner we keep up a constant current, flowing on top to the automatic skimmer and filter, and another reverse current of the filtered juice returning by way of the bottom of the pan, to again come in contact with the heat and thrown to the top, separating the remaining impurities and keeping up a constant circulation of the juice and producing the most rapid evaporation that can be made, and the strainer and filter catching and retaining all the impurities of the minutest character that have been separated from the juice, and preventing them from again mingling with the boiling juice and giving it a bad flavor and darker and cloudy appearance. All experts in the use of steam concede that in order to produce the most rapid evaporation there must be a constant circulation, and we are very much gratified with the manner in which our pans have operated, as they have enabled us to produce an article of sirup that has sold in the wholesale markets in competition with the best products of the country, made by either the open pan train or vacuum-pan and char-filters combined.

It saves labor, and above all things we prize it on account of its perfect work skimming the juice, and not endangering a depreciation in the value of the sirup by being imperfectly skimmed by tired and careless help; for without perfect skimming off of the impurities after they have once been separated, to keep them from being reboiled into the sirup again, there is danger that more or less of the batches or strikes will be run into the storage tank in a cloudy condition, and consequently of bad flavor, and help to destroy or depreciate what good sirup there is in the tank; and if it is intended for sugar it will be what is called a gray sugar, having a dull, dirty appearance. It was a case of this kind that occurred to us when we first commenced that suggested this plan of evaporation to me for our own safety and protection.

Our finishing pan is similar to our evaporator, but smaller in dimensions.

Our cooler works admirably, and is actually necessary in large works to cool the sirup immediately after finishing for commercial use or for sugar-making.

Our whole outfit, including land, buildings, and machinery, cost about $6,000, and has a capacity of making from eight to twelve hundred gallons of sirup per day. The amount of sirup made per day depends mostly upon the strength of the juice we are making.

In regard to my ideas of the future of this industry, I would say I have had no occasion to change my opinion expressed three years ago. I then made up my mind that if the industry was conducted on strictly business principles there was money in it for the farmer and the manufacturer of sirup alone, even if they should fail to produce sugar; and my past experience has confirmed that belief. And your own valuable experiments made at the university farm this past season, with the able assistance of the department chemist, Mr. Swenson, will dispel the doubts that existed in the minds of many, who could not possibly be persuaded to believe that sugar could be produced here at home, grown on our own farms.

The many central works and refineries devoted exclusively to the sorgo industry, that have been put in operation in many of the States, at a cost of from $5,000 to $60,000 each, is evidence of the fact that the most timid and skeptical factor in the development of this new industry—capital—has become convinced that it is a safe investment; after the most careful and searching scrutiny have united with science and skill and are partly carrying out the idea of central works, that I have been laboring to establish in this State, and the fine results you have obtained in your experiments will hasten the time of its realization.

There seems to me to be no other practical way of meeting the requirements of this rapidly growing business than by establishing central works—a central works located at some point accessible by rail from several directions, to facilitate receiving raw sirups from a large amount of territory, and fully equipped with all the latest improved mechanical appliances that have been tested and proven to be well adapted to the manipulation of the sorgo juice, to manufacture a first-class commercial sirup, and a soft white and yellow sugar. The central works should have a capacity of grinding from three hundred to five hundred acres of cane annually, to insure having a sufficient amount of business early in the season, so as to keep the works in operation as much as possible during the year. The central works could have nearly or quite all of their crop worked up before they would be able to obtain semi-sirup from the auxiliary works, for making sugar and refined sirups from. The central works should be under the management of some one who has a practical knowledge, and is qualified to instruct operators of the auxiliary works how to make the semi-sirup and leave it in proper condition for the central works.

Suitable buildings and machinery to work up 500 acres of cane, and rework the semisyrup made by the auxiliary works, from 3,500 acres, into sugars and sirups, taking eight months in the year, would cost $25,000, all fitted up ready for business.

It is not practicable to haul the cane more than three miles to mill, and to obtain a sufficient amount of raw sirups for a central works of such a character requires many auxiliary works, large and small, operated by steam or open fire train (steam being the cheapest and best, and destroys less sucrose, is preferable) to make the semi-sirup, which an intelligent and careful operator can do successfully by working under instructions from a competent manager of a central works.

To fit up a steam train so all the machinery will be properly proportioned, to insure the least expense in manufacturing, and produce an acceptable article requires the aid of some one who has sufficient practical knowledge to determine, when informed of the number of acres designed to be worked, the size of mill required, the amount of steam-generating power required, beyond the motive power, to evaporate the amount of juice expressed by the mill in less than an hour, and the number of square feet of heating surface it takes, with a given quantity of steam under a certain pressure, to evaporate the juice of a minimum strength down to semi-sirup, in the required time to produce the best results.

The lack of knowledge on these very essential points has been the means of causing some losses and discouragements to the owners of the works, and the growers of the cane also.

In conclusion, I beg leave to say in behalf of many farmers who have raised the cane, and many more who desire to do so, that I have conversed with on this subject in many different parts of our State, that they hope our representatives at Madison will realize what great interest it will be to the farmers and to the wealth of the State for them to make a special appropriation sufficient to enable you and your very able assistant, Mr. Swenson, the department chemist, to continue the valuable experiments you have commenced and that have produced such splendid results as to justify the belief that this new and valuable crop will be extensively raised by the farmers of this State in the near future.

They feel they have a right to ask for an appropriation for their agricultural department to make intelligent and systematic experiments (which the farmers are unable to do) to determine for them the best soils, fertilizers, &c., to use in developing for them a crop that gives such good promise of being of so great a value to them and the whole State. They also feel that they are behind the times in this matter, as other States have realized the importance of this crop to such an extent that they not only pay a premium on every pound of sugar that is made from the native cane raised in the State, and exempt from taxation for five years all the machinery employed in sugar making, but to encourage the farmers in growing cane they pay them a premium for every ton of cane they produce.

Hoping that you may be permitted to continue your experiments in this sugar ndustry with a sufficient amount of money at your disposal to enable you to extend your field of usefulness in this and any other direction that will be of benefit to our farming community, I remain,

Respectfully, yours,

A. J. RUSSELL.

Prof. W. A. HENRY.

[From J. D. Sherwood, Green Lake County.]

DARTFORD, GREEN LAKE COUNTY, WIS., *December* 18, 1881.

DEAR SIR: In reply to your favor of 8th ultimo, would say that I rolled $347\frac{1}{2}$ tons, averaging 7° B., allowing on the basis of 50 per cent. of juice expressed $10\frac{1}{2}$ gallons to the ton, which basis has given about 100 gallons to the acre on clay and sandy loam soil. The highest yield was $6\frac{565}{1000}$ tons, testing 8° B., from one-third of an acre, raised by William McConnell, of this town, being at the rate of 238 gallons per acre, and the lowest yield nigh about 30 gallons to the acre, juice 3° B. Commenced September 9 on the above yield, the seed of which was ripe, but most of the after-working was dough to ripe. Most of the cane was planted after other work, and then it has paid better than anything else; but not as well as last year, owing to the peculiar season. The cost of working our crop of eight acres was ten days' work fitting ground; eight days' work planting and cultivating; five days' work thinning out; forty-five days' work stripping and cutting, and then only one-half of it stripped, as it was badly lodged; twenty-four days' work and team drawing one and a half miles; making ninety-two days' work for $70\frac{3}{4}$ tons, testing 7° B., which was worked at 20 cents per gallon, and also at the halves costing to manufacturers, including the 20 per cent. wear on outfit costing about $4,000, 14 cents per gallon, which is more than it will next, owing to being inexperienced in everything. But still the consumers are well pleased, saying that they cannot replace it from the grocery. Families are using five gallons where they only used one before, with a very great difference in their sugar bill to their credit; and why not? It is cane sugar instead of the insipid glucose backed with a little sorghum that is dealt out by most of the stores as "sugar-house." There is no doubt at all in the fact that very soon we shall manufacture most of sugar and sirup and my very greatest fear is that it will be overdone, as those who raise it increase their acreage. I find that the best sales are made where it is known. It brings from 45 to 60 cents per gallon.

My outfit is a $3\frac{1}{2}$ Niles and complete steam train, with 12 horse-power engine and 45 horse-power boiler, from Blymer & Co., Cincinnati, Ohio. Burn bagasse and coal, which makes the costs about 5 cents per gallon.

Trusting that the above hastily condensed items are encouraging to you in your practical endeavors to place on a good foundation one of the best industries of the Northwest, and hoping that success will continue to crown your labors, I remain,

Very truly, yours,

J. D. SHERWOOD.

Professor HENRY, *Madison, Wis.*

[The following extract from a letter from Joseph H. Osborn, esq., Oshkosh, Wis., contains some valuable suggestions:]

I am satisfied that the sooner cane is worked up after it is cut, the better will be the character of the sirup made from it. I have no faith in the curing process which has been recommended frequently.

Again, the cane should be kept *clean*. Carelessness in this respect cannot be remedied in small works like mine. The dirt will be carried through into the sirup and is very damaging in its effect. Large establishments might provide for taking it out, but in this case prevention is better than cure.

I am very glad that the farmer and rural manufacturer is likely to have the aid of scientific gentlemen in developing this "new industry."

There are a great many things in connection with the manufacture of sirup the proper knowledge of which must come from a scientific source. Among these is the *correct* method of using the saccharometer. Scarcely a writer in the Rural World, upon the subject of Amber cane culture and manufacture, but refers to the test of the juice by the saccharometer. He may tell how he planted the seed and when; *how* he cared for the crop, and *how* he harvested it; but when he says the juice tested 7° B. or 12° B., he does *not* state what were the conditions of the test. Did he test the juice as it ran from the machine? If so, did he also test it by the *thermometer*? If it was not 60° by the thermometer, did he take means to make it so? If yea, *how* did he proceed?

Again, if he tested the juice by the saccharometer as it came directly from the mill, and also by the thermometer, even if the latter indicated 60°, did he allow the juice to stand an hour and test it again; and, if so, was the result the same? I think not; my experience is that there will be several degrees difference. If Professor Collier stated that juice tested a certain degree, I should of course know that the conditions of test were correct; but from my own experience I doubt very much if all the writers for the Rural World, who state results by the saccharometer, can be relied upon as having secured the correct conditions necessary for the test. It seems to me that correct information upon the correct use of the saccharometer should be given in a popular way for the benefit of those engaged in this Amber cane business.

Again, in regard to the use of lime. Are we to accept it for a settled fact that if the *cold* juice is tested with lime that it can be allowed to stand without injury for a length of time. (If so, how long?) If I remember correctly, this statement was made by Professor [?], of Illinois, through the Rural World.

Again, granted that lime is used with the cold juice, and heat subsequently applied to aid defecation, should the evaporation be proceeded with at once, or could the warm juice be allowed to stand any length of time; and if so, would it aid the clarification, or should or could some additional method of clarification be used before commencing the evaporation?

Again, in years gone by, when the making of sugar from corn-stalks was talked about, the removal of the young ears of corn was said to be essential to develop the greatest amount of sugar in the stalk. Question: Would science consider that the removal of the young seed tuft from the cane would add to the strength of the cane juice? Your circular called for facts. I have given mostly suggestions, or at at least I hope you will consider and treat them as such.

Truly yours,

JOSEPH H. ORTON.

[To those in doubt as to whether it pays to grow cane, I would refer the following letter sent me by one of our careful farmers. It is the most complete statement I have yet seen, and deserves careful attention:]

KENOSHA, WIS., *February* 26, 1881.

DEAR SIR: I herewith give you the result of growing one acre of amber sugar-cane in 1880. The plot of ground is composed of black muck, verging into a sand loam, two-thirds of the plot being the former and one-third the latter. There were about four rods of very low ground on which the cane grew very rank and lodged. There was no waste ground. In 1879 it was heavily manured and a very heavy growth of drilled fodder corn raised and plowed that fall. The ground was dragged and marked in rows one way, 3½ feet apart, extending north and south, on May 20, and on May 21 it was planted by hand. dropping the seed in the marks made by the marker and covering with the foot. Two pounds of seed was used. One-half of it was planted from 12 to 18 inches apart and the other from 12 to 25 inches. I think it would average seven or eight seed to a hill. It was then rolled and cultivated twice with a two-horse cultivator. One man spent one day on the piece with the hoe, cutting out grass between the hills. This would not have been necessary had the seed come up evenly. One-third of the piece was dry, and the seed not being covered any deeper, did not come up for two weeks, hence could not cultivate it evenly. It was stripped by hand at intervals from September 14 to September 27, cut and bound September 28, drawn to mill on the 29th and 30th, carefully weighed and piled. Total weight, 13⅐⅘ tons.

The first half, or that planted the thickest, weighed about 8 tons, and the other half 5½⅔⅔ tons. The cane was made up October 7 and yielded 170 gallons of sirup, weighing 11½ pounds to the gallon. The juice tested 7¾ by the saccharometer and was boiled down to 40. There was one load of leaves saved for fodder and three double boxes of seed which was fed to the pigs. I estimate the value of the crop as follows:

Dr.		Cr.	
To interest on land..............................	$2 00	By fodder..	$10 00
To half day's work plowing	1 50	By 170 gallons sirup at 50 cents	85 00
To dragging and marking.........................	50		
To 2 pounds seed................................	70		95 00
To planting....................................	1 00		
To hoeing	1 00		
To cultivating	1 00		
To stripping	6 00		
To cutting and binding..........................	3 00		
To topping and hauling..........................	10 30		
To hauling fodder and feed......................	1 00		
To four barrels at 75 cents	3 00		
To making 170 gallons at 20 cents...............	34 00		
	65 00		
By balance.....................................	30 00		
	95 00		

M. O. MYRICK.

Prof. W. A. HENRY, *Madison, Wisconsin.*

[The following letter from H. W. Small & Co., Chicago, will certainly be read with interest. It should be remembered that from the peculiar line of business of this company—that of supplying the wholesale trade with sirups and molasses—it is in a position unequaled by any other company in the west to judge upon the true merits of the case:]

CHICAGO, *December* 28, 1881.

DEAR SIR: We have your favor of 24th, with samples of sugar and sirup before us. You have obtained a remarkable yield from your experimental one-fifth acre. One thousand pounds of good brown sugar and 80 gallons of sirup per acre would be a very profitable crop for any of our farmers, and we read with very much interest your statement that the analysis of the cane showed nearly twice the quantity of sugar that you obtained; or, in other words, that the processes for extracting the sugar from Amber cane is so imperfectly understood at present, even by our most scientific men, that nearly one-half the yield is lost. Well, this only confirms our opinion the more strongly that the profitable raising of Amber cane in the north for the manufacture of sugar and sirup or molasses is no longer an experiment, but an assured fact; and, although but just in its infancy, enough has been already done to show that skill in its manufacture is the one great requirement.

Now, we not only would not advise every farmer to rush in blindly and plant a few quarter acres of Amber cane, but we would advise that they do no such thing until you who are giving so much time and attention to this business learn how, and "write a book" of instructions, so that every farmer may know how without the possibility of a failure. Then "exit" New Orleans, "enter" Amber.

We have received samples of Amber molasses this season that compare favorably with "New Orleans," while other lots have been very poor; and the difference, so far as we can learn, was not so much in the soil, or climate, or seed, as in the "modus operandi" of manufacture.

The sugar is there; the molasses is there. How to secure it, after it is grown and ready for the mill, is the one great question for you scientific men. We sincerely hope that the State will continue to foster this industry until it is thoroughly understood, so that every farmer can grow his own sugar and molasses at one quarter the present price of New Orleans, and, what may be even better than that, know that they have an absolutely pure article

The better grades of amber are slowly overcoming the old prejudice against sorghum, and we believe the time not far distant when a choice Amber molasses will be more sought after than a somewhat doubtful mixture of New Orleans glucose and sirup.

Wishing you every success, we are
Yours truly,

H. W. SMALL & CO.

Prof. W. A. HENRY, *Madison.*

  *   *   *   *   *   *   *

10.— *REPORT ON THE MANUFACTURE OF SUGAR, SIRUP, AND GLUCOSE FROM SORGHUM.*

[Based upon experiments made in 1880 and 1881, at the Illinois Industrial University.]

By HENRY H. WEBER, Ph. D., *Professor of Chemistry*, and MELVILLE A. SCOVELL, M. S., *Professor of Agricultural Chemistry.*

SIR: The undersigned have the honor to submit herewith their complete report of experiments in the manufacture of sugar from sorghum, made at the Illinois Industrial University during the seasons of 1880 and 1881.

Very respectfully,

H. A. WEBER, *Professor Chemistry.*
M. A. SCOVELL, *Professor Agricultural Chemistry.*

S. H. PEABODY, LL.D.,
 Regent Illinois Industrial University.

INTRODUCTION.

The object of the investigations made upon sorghum cane at the Illinois Industrial University was to settle, if possible, the much disputed question whether sugar could be made from this plant on a manufacturing scale and with commercial success. From the many conflicting reports relating to this matter no definite conclusions could be drawn, and it was found necessary, in order to prosecute our work in an intelligent manner, to treat the whole subject as an entirely new field of investigation. It has been claimed by many that the proper sphere of the sorghum industry is the production of sirup, and a great deal of good work has been accomplished in improving the quality and yield of this article. But what may have been true for sorghum a few years ago does not hold good to-day. The sorghum industry is at the present time confronted by another, namely, the glucose industry, which, although still in its infancy, has already shown its superiority in the production of sirup both in regard to quality and quantity. This statement is made with due consideration of the many attacks which the glucose industry has of late received. Glucose as an article of food is equal to if not superior to cane sugar, and its artificial production from corn or other amylaceous substances is a perfectly legitimate business. It is true that in the decolorization of the glucose injurious substances may be employed, and if the products sent to market are not perfectly free from them great injury may be done to the consumers. The same thing may be said for the refining of cane sugar. But in either case the employment of injurious substances is not a necessity, and should be condemned by every one who is interested in public welfare. Glucose, when made as it should be, is perfectly harmless, and no valid objection can be made to it in a sanitary point of view, when employed for any legitimate purpose to which it is adapted. The sorghum industry must regard the manufacture of glucose as a fair competitor, and the latter will never lose in importance by any unjustifiable attacks or criticisms. From these considerations it seems evident that the production of sirup alone can no longer maintain the cultivation of sorghum on a scale which would suffice to give it the name of an industry.

To accomplish this sorghum growers should turn all their attention and energy to the production of crystallizable sugar, which glucose, on account of its inherent properties, can never replace, and which will always find a ready market free from all competition.

These circumstances led to the investigations about to be described, and the results obtained have exceeded our most sanguine expectations. Our experiments, both scientific and practical, have shown beyond a doubt not only that the manufacture of sugar from sorghum in our own State is practicable, but also that it will be highly remunerative when undertaken on a large scale.

Up to the present time sorghum seed has never found a proper utilization. Although in its general composition it resembles other grain, as corn, the amount of tannin contained in it, as our analysis given farther on shows, will no doubt prevent its liberal use as food for animals. Knowing that immense quantities of seed will necessarily be produced as soon as the sorghum sugar industry is introduced, we have given this matter careful study, and have found that the seed is eminently adapted for the production of glucose. We have prepared the glucose directly from the ground seed, without the tedious and expensive process of first separating the starch. The great advantage of this industry to the sorghum industry will appear from the fact that as the seed is practically ripe when the cane is cut it can be stored up until the sugar season is over, and can afterward be manufactured into glucose with the same machinery now used in making sugar from the cane, thus giving employment for the balance of the year to the works, which otherwise would have to lie idle for eight or ten months annually.

Our work occupied two distinct fields of experiments: First, scientific investigations, in which the nature of sorghum cane was studied; second, practical experiments in making sugar.

PERIODICAL EXAMINATION OF THE CANES FOR SUGAR.

The objects of these analyses were:

1. To note the development and changes of the sugars in the plant during its growth.

2. To notice the changes which the cane undergoes after reaching this maximum stage in the quality and quantity of its saccharine matter: First, while standing in the field untouched; second, standing stripped two weeks; third, cut and lying under shelter.

3. To ascertain the portion of the stock richest in sugar.

4. To study the effect of different varieties of soils on the development of sugar in the cane.

5. To determine the effect of freshly manured soils on the development of sugar in sorghum.

6. To compare the different varieties of sorghum as sugar-producing plants.

These examinations were conducted in the following manner:

On the dates specified, ten average stalks were selected from the given field, stripped, topped just below the uppermost leaf, and cut off one joint above ground. The stripped and topped cane was crushed in a thoroughly cleansed Victor mill. The juice was collected in a bottle, and after being cooled down to 20° c., the sp. gr. was noted, then 10 c. c. were put into a graduated cylinder for the estimation of grape sugar, and 10 c. c. were put in a beaker for determining the amount of cane sugar.

For the estimation of grape sugar the 10 c. c. measured off for this purpose were diluted so as to measure exactly 100 c.c. and the grape sugar then determined by Fehling's solution.

The portion reserved for cane sugar was diluted, 12 drops of dilute sulphuric acid added, and the whole heated over a water bath for one hour. The mixture was then allowed to cool, sodium hydroxide added to alkaline reaction, diluted to 500 c.c., and the total amount of sugar determined with Fehling's solution. The difference between the grape and total sugar was estimated as cane sugar by multiplying by 0.95.

The results of the analyses are given in the tables which follow:

Table showing the development and change of sugars in sorghum.

Stage of development.	No.	Date.	Variety.	Specific gravity of juice.	Grape sugar.	Cane sugar.	Average of cane sugar.
Beginning to head	1	Aug. 14, 1880	Orange	1.055	5.70	4.90	4.14
	2	Aug. 10, 1881	Amber	1.058	8.39	3.38	
In blossom	3	Aug. 25, 1880	Orange	1.062	6.10	7.12	7.77
	4	Aug. 10, 1881	Amber	1.066	5.43	8.42	
Seed soft and milky	5	Aug. 14, 1880	Amber	1.065	3.34	10.75	
	6	Sept. 6, 1880	Orange	1.068	5.00	9.13	
	7	Aug. 10, 1881	Amber	1.068	4.25	9.84	
	8	Aug. 12, 1881	do	1.070	3.75	12.75	8.56
	9	Sept. 1, 1881	Orange,	1.048	6.11	3.71	
	10	Sept. 2, 1881	Orange	1.048	6.58	5.19	
Seed in hardening dough	11	Aug. 25, 1880	Amber	1.068	2.47	12.48	
	12	Sept. 16, 1880	Orange	1.065	4.11	9.76	
	13	Aug. 10, 1881	Amber	1.074	3.65	10.10	
	14	Aug. 12, 1881	do	1.074	2.65	1.337	
	15	Aug. 16, 1881	do	1.070	3.92	1.189	
	16	Aug. 16, 1881	do	1.072	3.00	1.366	
	17	Aug. 19, 1881	do	1.067	3.46	12.49	
	18	Aug. 19, 1881	do	1.074	3.10	13.18	11.95
	19	Aug. 19, 1881	do	1.076	2.97	13.64	
	20	Aug. 19, 1881	do	1.070	2.98	12.80	
	21	Aug. 19, 1881	do	1.070	3.26	12.52	
	22	Sept. 1, 1881	Liberian	1.060	3.67	10.24	
	23	Sept. 1, 1881	Amber	1.063	2.61	13.47	
	24	Sept. 1, 1881	do	1.056	2.18	11.14	
	25	Sept. 1, 1881	Chinese	1.052	4.13	8.60	
Seed ripe	26	Sept. 6, 1880	Amber	1.064	2.13	11.42	
	27	Sept. 16, 1880	do	1.065	2.79	11.02	
	28	Oct. 2, 1880	do	1.069	2.47	10.06	
	29	Oct. 6, 1880	Orange	1.078	4.02	11.41	11.18
	30	Sept. 9, 1881	I. I. U.	1.070	2.93	12.48	
	31	Sept. 1, 1881	Amber	1.070	2.71	10.77	
	32	Sept. 2, 1881	do	1.070	2.61	10.57	
	33	Sept. 5, 1881	do	1.067	3.16	11.76	

The analyses made in 1880, numbers 1, 3, 5, 6, 11, 12, 26, 27, 28, and 29, were from cane grown upon the University farm.

The following data in regard to the planting and cultivation of the cane were furnished by G E. Morrow, professor of agriculture:

Two varieties, Orange and Early Amber; seed obtained from Hedges, Saint Louis; planted by hand, May 14, 1880. The Orange was planted in a plot of nearly one acre (.955) in 24 rows four feet apart, in hills about four feet in a row. The Early Amber was planted in a plot of one and one-half acres (1.48) in 40 rows three and one-half feet apart, and with hills about same distance apart. Each plot was on good prairie soil which had been in corn two years, following a liberal application of barn-yard manure. The plots received ordinary field culture—a two-horse corn cultivator—except hand hoeing and thinning to four or five stalks when ten to twelve inches high. The suckers were not removed. The Orange averaged about seven feet in height, and over an inch in diameter at base. The Early Amber averaged over nine feet in height, and rather less than three-quarters of an each in diameter at base. The canes were cut about six inches from the ground. Of the Orange, from two to three feet of the top was taken off; of the Early Amber, rather more than three feet.

An analysis was made of the soil on which these two varieties of cane grew, and also of its subsoil and of a virgin prairie soil adjoining.

The following table gives the result of these analyses. No. 1 was prairie soil, No. 2 the soil on which the cane grew, No. 3 its subsoil:

Soil.	No. 1.	No. 2.	No. 3.
Organic matter	1.9414	2.4880	3.7551
Silicic acid	0.0798	0.0617	0.0975
Sesquioxide of iron	1.8367	1.4517	1.2650
Alumina	1.4775	0.5700	1.7150
Manganese	0.1798	0.2200	....
Phosphate of lime	0.1683	0.2103	0.1152
Carbonate of lime	0.3835	0.5845	1.2515
Corbonate of magnesia	0.5244	0.6757	0.7140
Potash	0.0733	0.0785	0.0505
Soda	0.0177	0.0211	0.0970
Sulphuric acid	0.1403	0.1519	0.2137
Soluble matter found	6.8327	7.5134	9.2745
Organic matter	4.1150	6.0700	8.9549
Silicic acid	72.1765	68.7127	68.0224
Alumina with trace of iron	12.7143	12.0520	9.3156
Lime	0.5729	0.7721	0.6444
Magnesia	0.4893	0.4831	0.4836
Potash	3.0041	3.0331	2.4561
Soda	0.5120	0.6344	0.5664
Manganese	0.0093	0.0847	....
Phosphoric acid	0.1933	0.1553	0.2628
Insoluble matter found	92.7867	91.9974	90.7062
	99.6194	99.5108	99.9807

Analyses Nos. 2, 4, 7, and 13 were made from cane grown upon the farm of Mr. J. W. Cushman, two miles south of Urbana. The field on which this cane was planted had grown seven consecutive crops of sorghum, without manure. It was high prairie land sloping towards the south. Seed planted April 25.

The cane of Nos. 8 and 14 was grown about one and one-half miles northeast of Urbana, on timber land. The field had been used as a barn-yard previous to its being planted with cane, and was therefore richly manured. The seed came from Minnesota through Mr. Le Duc, ex-Commissioner of Agriculture. The seed was planted the first week in May. Cultivated as usual for corn.

Results Nos. 15 and 16 were obtained from cane grown three miles south of Champaign, on virgin prairie. Eight rows were planted along the roadside, bounded on the outer side by the road itself and the inner by a tall, dense hedge-fence. Mr. Holmes, the owner of the cane, said the seed came from Mississippi and was planted the last week in April; land gradually rising from a slough near by. Two varieties of heads were present in this cane; the panicles of one (analysis No. 15) were clustered and erect; those of the other (No. 16) were spreading with pedicels drooping.

No. 21. University farm; volunteer cane, from cane grown on the field last year.

The cane from which analyses Nos. 17, 18, 19, and 20 were made was grown upon timber land about three miles northeast of Urbana. The seed probably came from Minnesota.

No. 17. Cane grown by Mr. E. Bishop; field ten years in cultivation, manured three or four years ago; seed planted about the middle of May; rows 3½ feet apart in hills 3 feet apart; an average of eight stalks on a hill; cane small; Nos. 18 and 19, cane grown by Christ. Shuman. No. 18 was on high land, twelve years in cultivation and had never been manured; an average of five stalks in a hill; growth of cane medium. No. 19 was on low land, four years in cultivation; average of eight stalks in a hill; cane large and thrifty.

No. 20. Cane grown by Samuel Wilson, on land four years in cultivation; hills 3 by 3½ feet apart; an average of eight stalks in a hill; field on the top of a small hill.

Analyses Nos. 9, 10, 22, 31, and 32 were made in Macoupin County, Illinois, Nos. 9, 22, and 31 from cane raised about two miles north of Virden, by Mr. Charles Rauch, and Nos. 10 and 32 one mile west of Girard, by Mr. D. C. Ashbaugh. The prairie soil in this county is very black, deep, and "mucky" No. 9, cane grown on timber land; seed planted May 12, 1881; hills 3 by 3, an average of five stalks in a hill. No. 22, volunteer cane; prairie land. No. 31, prairie land; seed planted first part of May. No. 32, prairie land; seed planted latter part of May.

, The results of experiment No. 53 were obtained from cane grown by Christ. Lust, about a mile west of Monticello, Piatt County. The field was timber land—a poor, clayey soil; seed planted first week in May.

Analyses Nos. 23, 24, and 25 were made of the juice of sorghum grown upon the so-called Mississippi sand-lands near Oquawka, Illinois. No. 23 was from cane grown by Dr. Park, one mile east of Oquawka. Nos. 24 and 25 were made from cane grown by Tom Ricketts, two miles northeast of same place.

Development of sugar.—Analyses Nos. 5, 11, 26, 27, and 28 were made from the same field on the date specified, and show conclusively that the cane sugar reached its maximum quantity when the seed was in the "hardening dough," and that it afterward gradually diminished. The same fact appears on comparing the average under each division in the table.

Effect of stripping and allowing to stand.—On October 2, 1880, an analysis was made of the juice of cane which had been stripped on the 18th of September—the cane not otherwise disturbed—with the following result:

Specific gravity of juice _____ 1.074
Grape sugar _____per cent__ 1.82
Cane sugar _____per cent__ 13.11

This subject needs further investigation.

Change of sugar after cutting the cane.—On October 23, 1880, an analysis was made of the juice of the Orange cane which had been cut, stripped, and topped October 2 and placed under shelter until examined. Juice whitish.

Specific gravity _____ 1.091
Grape sugar _____per cent__ 14.66
Cane sugar _____per cent__ 3.55

A sample of cane, cut August 25, 1880, without being stripped and topped, was preserved in a warm room where it had become dry long before it was examined. On April 3, 1881, it was analyzed and showed 12 per cent. of grape sugar and no trace of cane sugar.

Comparison of the upper and lower half of the cane.—The two following analyses were made to show what part of the cane is richest in sugar:

Amber—October 2, 1880.—Juice obtained from the upper half of the stalks after topping as usual.

Specific gravity _____ 1.069
Grape sugar_____per cent__ 2.94
Cane sugar_____per cent__ 9.67

Amber—October 2, 1880.—Juice obtained from the lower half of stalks.

Specific gavity_____ 1.070
Grape sugar_____per cent__ 1.94
Cane sugar_____per cent__ 11.64

Effects of soils.—The following analyses were made to study the effect of different varieties of soil upon the production of sugar in sorghum. But as other circumstances, as locality from which seed was obtained, time of planting, and manner of cultivation may affect the amount of sugar, many more investigations would have to be made before definite conclusions could be reached. The table, however, shows that sorghum can be grown successfully on all varieties of soil specified.

Table showing the effects of different soils on the development of sugar in sorghum:

Variety of soil.	Number.	Years in cultivation.	Fertilization.	Variety of cane.	Specific gravity of juice.	Grape sugar.	Cane sugar.	Average.
Prairie	1	27	Manured 3 years ago	Amber	1.068	2.47	12.48	
	2	7	No manure	Amber	1.074	3.65	10.10	Grape, 2.94.
	3	27	Manured 4 years ago	Amber	1.070	3.26	12.52	Cane, 11.28.
	4	Unknown	No manure	Amber	1.07	2.71	10.77	
	5	Very old	do	Amber	1.07	2.61	10.51	
Virgin prairie	6		No manure	Amber	1.07	3.92	11.89	Grape, 3.46.
	7		do	Amber	1.072	3.00	13.65	Cane, 12.77.
Timber land	8	Unknown	Barn-yard manure	Amber	1.074	2.65	13.37	
	9	10	Manured 4 years ago	Amber	1.067	3.46	12.49	
	10	12	No manure	Amber	1.074	3.10	13.18	Grape, 3.07.
	11	4	do	Amber	1.076	2.97	13.64	Cane, 12.87.
	12	4	do	Amber	1.07	2.98	12.80	
	13	Many	do	Amber	1.066	3.16	11.76	
Mississippi sand land.	14			Amber	1.063	2.61	13.47	Grape, 2.39.
	15			Amber	1.056	2.18	11.14	Cane, 12.3.

Effect of manure.—To ascertain the effect of manure a field was selected which had been used as a barn-yard for several years. A part of the cane was planted directly on the rotten manure pile. An analysis was made of a sample taken from this part of the field, as well as of a part away from the manure pile. The seed in each case was in the "hardening dough." The following is the result of the analysis:

Manured—Sp. gr. 1.063.   Grape sugar 2.65.   Cane sugar 10.89.
Unmanured—Sp. gr. 1.074.   Grape sugar 2.65.   Cane sugar 13.37.

Variety of cane.—From the table it appears that the Amber is best adapted for the production of cane sugar. The Orange and Liberian can also be employed advantageously in the latter part of the season, as they mature later. Their yield is greater per acre, and this fact no doubt would compensate for the less proportion of cane sugar to grape sugar contained in them. Analysis No. 25 of the Chinese cane seems to indicate that it would be unfit for the production of crystallizable sugar.

PROXIMATE ANALYSIS OF SORGHUM CANE.

An average portion of the Orange cut at the same time, October 6, as that used in experiment 29, was reserved, with tops and leaves still remaining for the analysis.

The leaves and two feet of tops were removed, and cross-sections taken between each joint of the remainder of the stalks. The proximate principles were then determined according to the following scheme: The sections, as soon as cut, were weighed and then dried in a water oven, allowed to cool in the air, weighed, finally pulverized, and put in a stoppered bottle. Of the dried substance, ten grams were required for sugar, fiber, starch, gum, and vegetable acids; one gram for hygroscopic water and a sh; one gram for total albuminoids; five grams for oil. The gram of dried cane reserved for water and ash was heated in an oven at 110° C. until its weight was constant. It was then ignited and the ash weighed. The ten grams for the estimation of sugar, &c., were macerated with water in a mortar, the water decanted, and this process continued several times, the decanted liquids being filtered by Bunsen's method, and finally the residue was thrown on the filter and washed until the filtrate measured one liter. One hundred c. c. of this solution was evaporated nearly to dryness on a water bath, then the desiccation completed by passing a current of dry air upon the residue by means of an aspirator, the temperature of the substance ranging in the mean time between 90° and 100° C. The residue was then weighed, incinerated, and weight of ash noted.

Albuminoids.—Four hundred c. c. of the aqueous extract were evaporated to a sirup on the water-bath, calcined gypsum added, the whole then dried and the residue ignited with soda lime.

Five hundred c. c. of the aqueous extract were rapidly evaporated nearly to dryness, and the residue exhausted with alcohol of 87 per cent. by repeated boilings with fresh portions of the solvent as long as it was colored. The liquids were filtered, the residue thrown upon the filter and washed with hot alcohol, and the washings added to the filtrate. Water was added to the filtrate, the alcohol expelled by heat, and then the solution diluted to 200 c. c.

Grape sugar.—One hundred c. c. of this solution were reserved for the estimation of grape sugar. The remainder was acidulated with dilute sulphuric acid, and boiled to convert the cane into grape sugar.

Cane sugar.—The cane sugar was then estimated with Fehling's solution, as usual.

Gum and vegetable acids.—The residue insoluble in alcohol was dried at 100 C., weighed, and then incinerated. This ash and the soluble albuminoids were subtracted from the total amount of residue, and the remainder estimated as gum and vegetable acids.

The residue left after extracting the ten grams of cane with water was washed with alcohol acidulated with sulphuric acid to dissolve the albuminoids, transferred to a beaker, and diluted to 200 c. c. Five c. c. of normal sulphuric acid were added, and the whole boiled for an hour on the water-bath, then filtered through Bunsen's filter. The filter was also cut into shreds and boiled with water containing 1 per cent. of sulphuric acid to dissolve any starch remaining on it. After filtering, the two filtrates were added, and the starch estimated from an alliquot portion by conversion into glucose.

The method was as follows: The starch solution was diluted to 500 c. c. Three separate portions of 50 c. c. each were transferred to prescription bottles, 10 c. c. normal acid added. The bottles were then stoppered with rubber stoppers firmly tied, and placed in a salt-bath and boiled respectively for three, four, and six hours. The contents of the bottles were then neutralized, diluted, and starch calculated from the amount of grape sugar present. The solution boiled six hours had 0.02 per cent. more starch than that boiled four hours. Three hours' boiling did not convert all of the starch into grape sugar. The residue from which the starch was taken was boiled with sodium hydroxide, thrown upon a weighed filter and repeatedly washed with the same solution, then washed with hot water, and finally with alcohol and then with ether. The washed residue was dried at 110° C. and weighed, then incinerated, the weight of ash subtracted from the former weight, and the difference estimated as fiber. The gram reserved for the albuminoids was ignited with soda-lime, and albuminoids determined as usual.

The oil was extracted by ether from 5 grams of the dried cane.

The total water was estimated by adding the per cent. of loss of the air-dried cane and the hygroscopic water.

RESULTS.

Composition of stalks of Orange cane in 100 parts:

Water	76.58
Grape sugar	3.00
Cane sugar	9.77
Starch	4.12
Fiber	4.54
Oil	0.07
Gums and vegetable acids	0.24
Soluble albuminoids	0.23
Insoluble	0.16
Soluble ash	0.68
Insoluble ash	0.06
	99.45

ASH.

The ash from the remaining dried cane was analyzed by the following method: The cane was incinerated at a low heat, pulverized, dried, and put in a stoppered bottle.

Chlorine.—Two grams of the ash were exhausted with water, silver-nitrate added to the extract and the whole acidified with nitric acid. The precipitate of chloride of silver was collected upon a filter, dried, ignited, weighed, and the chlorine calculated in the usual manner. The filtrate was treated with excess of hydrochloric acid, silver chloride removed, and the solution preserved.

Silica.—The ash insoluble in water was treated with hydrochloric acid, brought to dryness, moistened with hydrochloric acid, water added, and the residue thrown on a weighed filter. The filter and its contents were heated at 160° C. until of constant weight, then ignited, and the silica weighed. The loss found between the two weights was called charcoal.

The solution from which the chlorine had been precipitated and the filtrate from the silica were mixed, and the whole diluted to 200 c. c. and well shaken. Fifty c. c. of this solution were reserved for the estimation of sulphuric acid and alkalies, 50 c. c. for phosphoric acid, manganese, lime, and magnesia.

Iron.—The remaining 100 c. c. were treated with sulphuric acid, and heated upon a water bath until the chlorine was expelled; then transferred to a flask, water and sulphuric acid added, and the iron reduced with hydrogen, generated by zinc suspended in the liquid by means of a platinum wire. To facilitate the operation, a strip of platinum was introduced into the flask and allowed to come in contact with the zinc. After the reduction the iron was estimated by a standard solution of potassium permanganate.

Phosphoric acid.—A solution of ferric chloride was added to the portion reserved for phosphoric acid, &c., in sufficient quantity for the iron to combine with all the phosphoric acid present. Sodium carbonate was added until the last drop caused a precipitate, which did not redissolve upon agitation. The mixture was then heated, a hot solution of sodium acetate added, and the whole brought to the boiling temperature, filtered, and washed with hot water.

The residue was dissolved in nitric acid and concentrated to about 10 c.c.; a nitric acid solution of molybdate of ammonia was added in excess, and the mixture allowed to stand in a warm place for twenty-four hours. The precipitate was collected on a filter, the beaker rinsed, and the contents of the filter washed with a mixture of the molybdate solution and water. The precipitate was dissolved in the smallest quantity of ammonia. Any of the phospho-molybdate precipitate remaining in the beaker was dissolved in a mixture containing three parts of water and 1 of ammonia and thrown upon the filter; finally, the filter was washed with the ammoniacal water. The filtrate boiled, and the phosphoric acid precipitated with a mixture of ammonium-chloride, magnesium sulphate and ammonia, made according to Fresenius's formula. After allowing the mixture to stand twelve hours, the precipitate was collected on a filter, washed with ammonia water, and the volume of the filtrate and washings noted.

The precipitate was ignited in a platinum crucible, a little nitric acid added, and again ignited to oxidize the charred matter present, cooled, and weighed. As ammonia-magnesia-phosphate is soluble in about 54,000 parts of ammoniacal water, .003 of a gram was add to this weight, as the filtrate measured a little over 150 c.c. The phosphoric acid was then calculated from this weight of pyrophosphate of magnesium.

Manganese.—The solution from which the iron and phosphoric were precipitated was treated with a few drops of bromine, and boiled to precipitate the manganese. The precipitate was collected upon a filter and thoroughly washed, then strongly ignited, and weighed.

Lime.—The above filtrate was concentrated, and while hot a little ammonia added, and then in excess of ammonium oxalate, to precipitate the lime. The mixture was allowed to stand twelve hours. The precipitate was then collected upon a filter, washed, dried, and ignited in a platinum crucible. After the filter was reduced to ash, carbonic acid was passed over the ignited lime to reconvert any oxide formed into carbonate. From the weight of calcium-carbonate thus obtained the per cent. of lime was calculated.

Magnesia.—The filtrate from the lime was concentrated, ammonia added in excess, and then a solution of phosphate of soda to precipitate the magnesia present. This precipitate and its filtrate were treated the same as the corresponding one, the estimation of phosphoric acid. The magnesia was calculated from the amount of pyrophosphate of magnesia found.

Sulphuric acid.—The 50 c.c. of the solution reserved for this purpose were boiled, and the sulphuric acid precipitated, with a slight exces of barium-chloride. The precipitate was collected upon a filter, washed, ignited, and weighed.

Potassa.—The above solution was treated, after concentration on a water-bath, with ammonia and ammonium-carbonate as long as any precipitate was formed, digested on a water-bath, filtered, and the contents of the filter carefully washed. The filtrate and washings were evaporated to dryness on a water-bath, and the residue ignited to expel ammoniacal salts. This residue was then treated with five and one-half times its weight of pure oxalic acid in the form of a concentrated solution, then evaporated to dryness, and again ignited to dull redness. The ignited residue was treated with a small quantity of boiling water, thrown upon a filter, washed with hot water, hydrochloric acid added to the filtrate, the mixture evaporated to dryness, and gently ignited, and the weight of the alkaline chlorides ascertained.

The separation of the alkalies was effected with platinic chloride, as follows:

The residue of alkalies was dissolved in a little water, and enough platinic chloride added to combine with the alkalies estimated as potassium salt. This mixture was evaporated nearly to dryness over a water-bath, care being taken not to boil the water. A mixture of six volumes of alcohol and one of ether was poured over the residue, and the whole allowed to stand several hours in a covered vessel, with occasional stirring. The insoluble potassio-platinic chloride was transferred to an equipoised filter, washed with alcohol and ether mixed, and finally dried at 100° C., and weighed.

Soda.—From the weight of the double potassium chloride the amount of the potassium chloride was ascertained. The weight was subtracted from the weight of the combined alkali chlorides, and the remainder called sodium chloride, and calculated as soda.

Carbonic acid.—One gram of the ash was transferred to a Rose carbonic acid apparatus, and the carbonic acid estimated by loss. The following were the results obtained :

COMPOSITION OF ASH.

Silica	27. 91
Iron oxide	0. 14
Phosphoric acid	5. 37
Manganese oxide	0. 89
Lime	6. 82
Magnesia	4. 64
Sulphuric acid	6. 23
Potassa	46. 48
Soda	0. 98
Sodium chloride	0. 42
	99. 88

ANALYSIS OF SORGHUM SEED.

A sufficient quantity of the seed was ground as fine as possible in an iron mortar, and was preserved in a glass-stoppered bottle.

The following portions of the ground seed were taken :

10 grams, for the estimation of sugar, dextrine, starch, and fiber.
1 gram, for the estimation of water and ash.
1 gram, for the estimation of albuminoids.
1 gram, for the estimation of oil.
1 gram, for the estimation of tannin.

Sugar, &c.—The ten grams reserved for sugar, &c., were rubbed up thoroughly with water in a mortar, then transferred to a filter and washed well with water.

Solution = A.
Residue = B.

The solution, A, was concentrated to about 10 c.c. in a porcelain dish on a water-bath, then transferred into a strong prescription bottle and washed with about 10 c.c. of water, and the washings added. Five c.c. of normal sulphuric acid were added, the bottle closed with a rubber stopper securely tied. The bottle and its contents were then transferred to a salt bath and boiled for six hours. After cooling, the contents of the bottle were transferred to a graduated cylinder, neutralized and diluted to 100 c.c., the coloring matter precipitated with acetate of lead, and, after thoroughly mixing, the whole was allowed to stand until the precipitate had settled to the bottom. A portion of the clear liquid was then transferred to a burette and dropped into 10 c.c. of Fehling's solution, diluted four times, and at the boiling temperature, until the whole of the copper had been precipitated as cuprous oxide. This point was determined by filtering a small quantity from time to time, acidifying the filtrate with acetic acid, and testing for copper with ferro-cyanide of potassium. The number of c.c. of the sugar solution it took was noted, and the sugar and dextrine determined by the following proportion:

1. The number of c.c. it took to precipitate copper solution: total number of c.c. : : .05 (grains of grape sugar required to precipitate 10 c.c. of Fehling's solution) : x.

X multiplied by 0.95 will give the grams of sugar in 10 grams of seed.

The residue, B, was washed on the filter with alcohol acidulated with sulphuric acid and finally with water, to dissolve the gluten. Then the residue was washed off the filter into a beaker diluted to about 400 c.c., 5 c.c. of sulphuric acid added, and the whole boiled on a water-bath until the liquid had no milky appearance. It was then filtered through an equipoised filter and washed.

Solution = C.
Residue = D.

Solution C was diluted to 500 c.c. Fifty c.c. of this solution were transferred to a prescription bottle and then treated as above for sugar and dextrine. From the grape sugar obtained the amount of starch was calculated.

Residue D was boiled with hot sodium-hydroxide, again thrown upon the filter and washed with the same solvent; afterwards with hot water, then with alcohol, and finally with ether. The washed residue was dried at 119° C., weighed, ignited, and the amount of ash deducted. The remainder was estimated as fiber.

Water.—For the estimation of water the ground seed was weighed in a glass-stoppered test-tube. After weighing, the glass stopper was replaced by a rubber one, through which passed two glass tubes, bent at right angles. One of these tubes was connected with an aspirator; the other with a calcium chloride tube and a sulphuric acid drying bott'e. The test-tube and its contents were then placed in an opening of a drying oven, who e temperature was between 100° and 110° C. During the operation a current of air, passing through the sulphuric acid and calcium chloride tube, thus drying it, was drawn into the tube and the moisture sucked out by means of the aspirator. When the weight became constant the loss was estimated as water.

Ash.—The contents of the tube were transferred to a platinum crucible, incinerated, and ash weighed.

Albuminoids.—One gram of the ground seed was ignited with soda lime. The substance was intimately mixed with a portion of soda-lime sufficient to fill a 14-inch combustion tube two-thirds full. About two inches of the tube were filled with soda-lime, then the mixture of soda-lime and substance added, the mortar rinsed with soda-lime, and finally the rinsings and enough soda-lime to nearly fill the tube. A plug of asbestus was put in, and the tube gently tapped to insure an air passage throughout its length.

Will's bulbs were charged with a deci-normal solution of oxalic acid. The tube, being placed in the combustion furnace, was connected with the bulbs. The fore part of the tube, containing the soda-lime only, was heated to redness, then heat applied, one jet at a time, along the entire length of the tube, care being taken that the combustion was completed in that portion of the tube where heat was applied before other jets were turned on, and also that the combustion was not too rapid. After the combustion was ended the contents of the bulbs were transferred to a beaker, tincture of litmus added, and the excess of acid titrated with a deci-normal solution of potassa. The amount of ammonia found to be present was calculated as nitrogen. The nitrogen was multiplied by 6.25 and the result called albuminoids.

Oil.—The one gram of ground seed reserved for the estimation of oil was placed in a short test-tube, the bottom of which was drawn out in the shape of a cone, with a small opening at the apex. A small filter placed in the cone kept any of the substance from passing through the opening. The tube was suspended in a small flask, and this stoppered with a cork through which a long glass tube passed. The whole was placed in a water-bath, ether (½ oz.) put in the outer tube, and heat applied to the water-bath until the temperature of the water boiled the ether. This operation was continued for half an hour. The percolate transferred to small weighed beaker, ether evaporated, and the beaker and its contents dried at 100° C., and then weighed.

Tannin.—One gram of the pulverized seed was digested with hot water for several hours, and the tannin estimated by a standard solution of gelatine.*

COMPOSITION OF SORGHUM SEED—ORANGE.

Sugar	0.56
Starch	63.09
Fibe	6.35
Water	12.51
Ash	0.64
Albuminoids	7.35
Oil	3.08
Tanr in	5.42
Total	99.00

EXPERIMENTS IN SUGAR MAKING.—1880.

The grinding of cane and the evaporation of the juice began on the 18th of September. It was the intention to begin working up the Early Amber as soon as possible after it had reached its maximum per cent. of cane sugar, and thus have it finished by the time the Orange was ready to harvest, leaving a small portion for subsequent experiments. Owing to the delay in the arrival of machinery the work was not begun until the above date.

The Early Amber had been ripe for over two weeks and was lying prostrate from the effects of a storm. The Orange was ripe. The object of these investigations was to see whether any method of manufacture of the juice into sirup could be depended upon to insure the subsequent crystallization of the sugar.

These investigations were undertaken with a view to the simplicity of machinery used and to the economical manufacture of the sirup, so that they could be of practical use to the farmer should any of the experiments prove successful.

The apparatus used for crushing and pressing the cane was a two-horse Victor mill,

---

*This remarkable result showing 5.42 per cent. of tannin requires confirmation.—*Committee.*

with three upright rollers. ˙The juice was evaporated in Cook's evaporator, with furnace attached, and of the size recommended for use with a two-horse crusher. The remaining apparatus consisted of barrels, tubs, pails, &c.

An attempt was made to heat the juice for skimming and clarification after it had been treated by chemicals, in the pan of a steam boiler of the form used by farmers to cook food for cattle. This boiler was found unfit for the purpose, as the temperature of the juice could not be raised in it above 108° C. A small pan was made, similar in construction to a Cook's evaporator, but furnished with a double bottom. The steam space in the bottom was about two inches high, and was connected with one of the boilers in the chemical laboratory. The object was to test the feasibility of evaporating the juice by steam under pressure with shallow pans.

In the experiments which follow the juice was either evaporated directly after it came from the mill, i. e., without the use of reagents, or after it had been submitted to clarifying processes. In the first the juice is designated in the experiments as *not clarified*, in the second as clarified, defecated, or *neutralized*.

THE EXPERIMENTS.

1. *Early Amber.*—September 18. Cane, very ripe and down; juice, *not clarified*—evaporated to a sirup which upon cooling weighed 11 pounds to the gallon. It was of a light color and had a distinct sorghum taste. Stalks, stripped and topped, yielded 48 per cent. of juice, having a specific gravity of 1.066. The sugar, not crystallized.

2. *Early Amber.*—September 20. Juice defecated. As the juice was brought from the mill milk of lime was added, little at a time, until a piece of red litmus paper would change to purple when dipped into the juice. Then a solution of tannic acid, and finally gelatine was added. The juice was then boiled and well skimmed and concentrated to sirup. The sirup was scorched and had a taste of extract of licorice. A small portion of the sirup evaporated to almost candy was readily crystallized.

3. *Early Amber.*—September 21. Juice not clarified. The evaporation was continued until the sirup upon cooling weighed 11 pounds. The sugar did not crystallize.

4. *Early Amber.*—September 22. Juice made alkaline with lime, and then neutralized with sulphate of alumina. Concentrated to a sirup that weighed when cooled between 11 and 11½ pounds. Sugar crystallized.

Before expressing the juice for this experiment the rollers were moved closer together and the cane crushed so much that the bagasse as it came out fell in pieces. Fifty-one per cent. of juice was obtained with a specific gravity of 1.068. One row of cane (0.037 acres) was taken for this experiment, producing 23 gallons juice, from which was made 3.17 gallons sirup, weighing 11¾ pounds per gallon. Calculating from this data, an acre of the Early Amber would yield 624.2 gallons of juice, or 86.1 gallons of sirup.

5. *Orange.*—September 23. Juice neutralized with milk of lime; afterwards tannin and gelatine added; evaporated to a sirup of 12 pounds to the gallon; sirup dark. The sugar commenced crystallizing in a few days. Three weeks afterwards the sugar was separated from the sirup by a centrifugal separator. Sugar brown.

In this experiment 360 pounds of topped and stripped stalks were used, producing 155 pounds of juice (43 per cent.); 28 pounds sirup (7.78 per cent. of the stalks and 18.04 per cent. of the juice); 13½ pounds sugar (3.8 per cent. of stalks, 8.87 per cent. of juice, 49.1 per cent. sirup).

One row (.0398 acres) yielded 30 pounds juice. Calculating the yield of an acre from these data we have 754 gallons juice, 120.6 gallons, or 1,447.2 pounds sirup, and 710.67 pounds sugar.

6. *Orange.*—September 24. Juice neutralized with lime, and a few drops of tannin added to every 10 gallons juice; then one-eighth ounce gelatine, and afterwards a little sulphate of alumina. Juice evaporated to a sirup of 11 pounds to the gallon; color very light. Sugar began crystallizing after standing two days.

7. *Orange.*—September 27. Juice neutralized with lime, and concentrated to a sirup of 11 to 12 pounds per gallon. Sugar readily crystallized.

8. *Orange.*—September 27. Juice neutralized with milk of lime; sulphurous acid was added to combine with any lime remaining uncombined in the juice. The sugar began crystallizing as the sirup was cold.

9. *Orange.*—October 1. Juice defecated with lime and sulphate of alumina. Sugar began crystallizing after three days. In this experiment stripped and topped stalks were used, yielding 54.2 per cent. of juice; specific gravity, 1.076.

10. *Orange.*—October 1. Juice evaporated without defecation. The sirup, after standing about five weeks, had but few crystals of sugar. In a subsequent analysis of this sirup (see analysis of sirup No. 4) there was found to be 38.9 per cent. of cane sugar and 26.91 per cent. of grape sugar.

1). *Orange.*—Juice not defecated; evaporated to a sirup of 12 pounds to the gallon. The sugar has not crystallized.

1:. *Amber.*—Juice defecated with lime and sulphate of alumina. The juice was quite acid as it came from the mill. Sirup black. Sugar crystallized.

F nding that some of the sirup whose juice had not been defecated did not crystallize, it was thought that perhaps a further concentration would cause the sugar to crystallize. For this purpose the sirup produced in experiment No. 3 was selected. In the early part of November it was further concentrated in the steam evaporator, but this had no effect upon the crystallization of the sugar.

F nding that the concentration of the sirup did not cause the sugar to crystallize, an analysis of several of the sirups was undertaken in order to investigate this subject more thoroughly. The following sirups were selected to be analyzed:

N). 1. *Early Amber.*—Sirup taken from that made in experiment No. 3.

N). 2. Sirup of No. 1 subjected to further concentration.

N). 3. *Orange.*—Sirup of experiment No. 9, with the crystallized sugar taken out by the centrifugal separator.

N). 4. *Orange.*—*Obtained from the sirup of experiment No. 10.* The following were the results obtained:

COMPOSITION OF SORGHUM SIRUPS.

Number.	Cane sugar.	Grape sugar.	Gum.	Water.	Ash.	Total.
No. 1	47.22	14.70	6.80	29.4	1.97	100.1
No. 2	45.62	20.00	10.51	20.39	3.78	100.3
No. 3	35.63	26.82	6.75	28.67	1.40	99.27
No. 4	38.9	26.91	7.80	21.04	1.75	96.40

The cause of the large per cent. of ash shown by No. 2 was undoubtedly the lime added to neutralize the sirup before the second concentration.

From the proximate analysis of the cane it appears that one acre of sorghum produces 2,55) pounds of cane sugar. Of this amount we obtained 710 pounds in the form of good brown sugar, and 265 pounds were left in the 737 pounds of molasses drained from the sugar. Hence 62 per cent. of the total amount of sugar was lost or changed during the process of manufacture. This shows that the method of manufacture in general use is very imperfect.

EXPERIMENTS IN SUGAR MAKING IN 1881.

Last year a large number of experiments were made in order to determine the means by which the cane sugar could be made to crystallize. This object was much more readily attained than we at first expected, and consequently we selected from those experiments the one which was most simple and most likely to be practiceable when operating on a large scale. In perfecting this our attention was given to the production of sugar and sirup which should be free from the objectionable sorghum taste and odor. In this we succeeded perfectly. Sorghum juice in its normal condition is acid. The conversion of cane sugar into grape sugar by boiling a solution of the same with a strong acid, as sulphuric or hydrochloric, has long been known to chemists. All other acids, even the weak organic acids contained in sorghum juice, act in a similar manner. Hence it will readily appear why in the ordinary manner of making sorghum sirup so little of the cane sugar originally contained in the juice can be made to crystallize. A great deal of the cane sugar is converted into grape sugar during the processes of defecation and evaporation, and what remains unchanged is prevented from granulating by the undue proportion of grape sugar produced. To avoid this loss of cane sugar we neutralize the juice when cold with calcium carbonate or milk of lime, or both. This part of the process requires skill and care, as the subsequent defecation of the juice depends upon it. After thus neutralizing the juice it is heated to boiling and thoroughly defecated. It is then passed through bone-black filters and finally evaporated to crystallization. The sugar and molasses obtained by this process are unobjectionable in regard to color and taste.

Experiment 1, August 22, 1881.—The cane selected for this experiment was grown on land which had previously been used as a barn-yard, the same as in analyses Nos. 8 and 14. The seed was nearly ripe and the cane very thrifty.

Weight of cane crushed _____ 1,560.00 pounds.

Weight of juice obtained _____ 687.50 pounds.

Per cent. of juice _____ 43.40

The juice was carefully neutralized with milk of lime, and brought to the boiling point in the defecating pan. A very heavy green scum rose, and this being removed the juice was seen to be full of a green, light flocculent precipitate, which did not subsequently rise to the top in any considerable quantity. The juice was now drawn off into tubs, where it was allowed to repose twelve hours. At the end of this time only about one-half of the juice could be drawn off clear, the precipitate being still suspended in the remainder. It was found impossible to filter this portion, and it was therefore thrown away. The clear juice, after being passed through bone-black, was evaporated in a copper finishing-pan to the crystallizing point. The melada had a very unpleasant saltish taste, owing to the presence of salts of ammonia. The sugar crystallized very readily, and although it looked well it still retained somewhat of this saltish taste after being separated from the molasses. Unquestionably this excessive amount of albuminoids—the green scum and suspended precipitate—was taken up by the plant from the nitrogenous elements of the manure, and the saltish taste was due to ammonium salts which came from the same source.

Manure therefore not only has a deleterious effect upon the development of sugar in cane, but it also prevents the thorough defecation of the juice which is necessary to the manufacture of sugar.

Experiment 2, August 25.—Cane same as that of which analyses Nos. 15 and 16 were made. Size of field, three-sixteenths of an acre.

CALCULATIONS FOR ONE ACRE.

	Pound
Stripped cane with tops	18, 535. 3
Stripped cane without tops	15, 765. 9
Weight of juice obtained	6, 545. 6
Per cent. of juice of stripped and topped cane	41. 52
Weight of melada from juice	1, 298. 7
Weight of melada from bagasse	253. 9
Total weight of melada	1, 552. 6
Weight of sugar from juice	504. 0
Weight of sugar from bagasse	104. 7
Total weight of sugar	608. 7
Weight of molasses from juice	794. 7
Weight of molasses from bagasse	149. 2
Total weight of molasses	943. 9

Calculations for one ton of topped and stripped cane:

Weight of juice	830. 4
Weight of sugar	77. 2
Weight of molasses	119. 7

To obtain the sugar from the bagasse it was packed in large barrels as it left the mill and was exhausted with water. The percolate thus obtained was treated like juice.

Experiment No. 3, September 17.—Early Amber; obtained from universty farm; volunteer growth among the corn; seed ripe; cane mostly blown down.

	Pounds.
Weight of stripped and topped cane	1, 440
Weight of juice	637
Per cent. of juice	44. 2
Weight of melada obtained	145. 8

Experiment No. 4.—Early Amber, grown upon university farm:

Weight of stripped and topped cane	1, 661. 0
Weight of juice obtained	603. 5
Per cent. of juice	36. 33
Weight of melada from juice	95. 5
Weight of melada from bagasse	13. 5
Sugar from juice	41. 5
Sugar from bagasse	6. 0
Molasses from juice	.54. 0
Molasses from bagasse	7. 5

In the last two experiments the cane was poorly developed and full of suckers, and consequently poorly adapted for the production of sugar.

GLUCOSE FROM SORGHUM SEED.

Our experiments have shown that as good glucose can be made from the seed of sorghum as from any other starchy substance. The yield of glucose or grape sugar is three-four hs or more of the weight of seed employed. The tannin does not interfere, as it is converted into glucose by the same means which are used to convert the starch, namely, boiling with dilute acids.

RECEIPTS AND EXPENSES OF ONE ACRE OF SORGHUM.

On the basis of the results actually obtained as described in the foregoing pages, we have calculated the receipts, and from the best data at hand the expenses, for one acre of sorghum.

Balance sheet.

RECEIPTS FROM SUGAR AND MOLASSES.

600 lbs. sugar at 7 cents	$42 00	
85 gallons molasses	34 00	
		$76 00

EXPENSES.

Cultivating one acre	$10 00	
Stripping and cutting	2 50	
Hauling	6 00	
Four days labor	6 00	
Fuel	1 00	
Barrels	4 00	
Freight and drayage	8 00	
		37 50
Net profit on sugar and molasses		$38 50

RECEIPTS FROM GLUCOSE.

1,250 lbs. glucose at 2 cents		$25 00

EXPENSES.

Gathering seed	$2 00	
Fuel	1 50	
Labor	1 00	
Barrels	4 00	
		$9 50
Net profit on glucose		15 50
Total net profit on one acre of sorghum		54 00

GENERAL CONCLUSIONS.

1. Seed should be planted as early as possible.
2. The proper time to begin cutting the cane for making sugar is when the seed is in the hardening dough.
3. The cane should be worked up as soon as possible after cutting. Cane which is cut in the afternoon or evening may safely be worked up the following morning.
4. The manufacture of sugar can be conducted properly only with improved apparatus and on a scale which would justify the erection of steam sugar-works, with vacuum-pans, steam defecators and evaporators, and the employment of a competent chemist to superintend the business. The same is true for the manufacture of glucose from the seed. Our experiments were made with the ordinary apparatus used in manufacturing sorghum sirup, and any person who desired to work on a small scale could use the methods with good results, provided he had acquired the necessary skill in neutralizing and defecating the juice and in the treatment of bone-black filters. The manufacture of glucose on a small scale is entirely out of the question. Five hundred to a thousand acres of sorghum would be sufficient to justify the erection of steam sugar-works, and this amount could easily be raised in almost any community within a radius of one or two miles from the works.

11.—*LETTER FROM H. A. WEBER TO PROFESSOR SILLIMAN.*

ILLINOIS INDUSTRIAL UNIVERSITY CHEMICAL DEPARTMENT,
Champaign, Ill., March 18, 1882.

SIR: Your circular asking for communications in regard to the "sorghum industry" is at hand.

Have sent you by to-day's mail a report of experiments in this line made by my colleague and myself, and hope that you may find something in it which will be of interest to you. In regard to the analytical as well as practical investigations it may be proper for me to state that they were our own personal work, and that we at least are satisfied of the correctness of the results as given.

A stock company has been formed here for the purpose of erecting steam sugar-works to test this matter on a commercial scale the coming season.

Yours, respectfully,

H. A. WEBER.

B. SILLIMAN, *New Haven, Conn.*

12.—*LETTERS FROM A. J. RUSSELL, JANESVILLE, WIS.*

[A. J. Russell, President of Wisconsin Amber Cane Growers' and Manufacturers' Association, and Chairman of Committee for the Purchase of Seed and Sugar Machinery. A. J. Russell, J. Boub. Office of Excelsior Amber Cane, Sirup, and Sugar Works, A. J. Russell & Co., Proprietors.]

JANESVILLE, WIS., *December* 28, 1881.

DEAR SIR: Your valued favor of the 26th at hand, and in reply to the several questions contained therein I would state that the yield of stripped stalks of cane per acre depends upon the quality of the seed, soil, and fertilizers used, method of planting, thoroughness of cultivation, and the season for growing the cane.

We have received at our works here and at other works I have been interested in from 7 to 20 tons per acre the "same season."

The yield of sirup per ton varies from 9 to 20 gallons according to the strength of the juice and the ability of the mill used in crushing the cane, so as to obtain the largest percentage of juice, and the economy of the mechanical appliances used in reducing the juice to proof or sugar density.

The yield of sugar, per ton, depends upon the amount of sucrose contained in the juice; the machinery used in reducing the juice to the required density, with the least destruction of the sugar, and the ability of the operator to handle the appliances used so as to separate and remove all the impurities that obstruct granulation, and varies from 7 to 9½ pounds per gallon of sirup.

Owing to the farmers having an imperfect knowledge of the proper care of cane, and a well-developed system of planting and fertilizing to obtain the best results, the average has only been 10 tons of cane per acre, and 14 gallons of sirup per ton, and 7½ pounds of sugar per gallon.

The sugar cost, to produce the cane and deliver at the mill, and manufacture and barrel it, ready for market, including cost of barrels, 2¾ cents per pound. From the experience of some of our most practical farmers in growing the cane, we are confident they can produce 20 tons per acre, in most of the seasons, and our experience in manufacturing determined the fact to our satisfaction that we can, in good corn-growing seasons, produce from cane grown on proper soil, where the required fertilizers have been used [produce], 17 gallons of sirup of sugar density per ton, and 9½ pounds of sugar per gallon, with the right kind of machinery, by taking out the first and second and possibly the third crop of crystals. The drainage syrup left will be a good commercial article, and if the farmers will save their seed, which they can do as cheaply as they can oats, they can obtain from 25 to 35 bushels of threshed seed per acre; and those who have fed it to their stock pronounce it of more value than oats as feed. If the manufacturer purchases the cane from the farmers by the ton, the cost of the sugar will be correspondingly greater, but not to exceed 3¼ cents per pound.

Respectfully yours,

A. J. RUSSELL,

To Hon. GEO. B. LORING,
Commissioner of Agriculture, Washington, D. C.

JANESVILLE, WIS., 3, 22, 1882.

DEAR SIR: In reply to your favor of the 13th would say, that I have obtained, with good machinery, 280 gallons of sirup per acre and 7½ pounds of sugar per gallon. Sugar sold for 9 and 10 cents per pound; sirup in job lots to wholesalers at 50 cents per gallon. The sugar was a very light yellow, polarized 96.6. The sirup was a very light straw color, transparent, and free from all sorgo or foreign flavor, and was placed in front rank, with New Orleans molasses (choice).

The above is from my own practical experience without any patented "process." The yield of seed is from 25 to 40 bushels per acre, and is used for all classes of stock, and sells at 50 cents per bushel. Data as to cost of raising the cane and delivered at the mill. 3 miles, which is as far as it is practical to haul. also cost of steam, or fire trains, that has produced the sugar and sirup that obtained the premiums at the convention; cost of manufacturing on either steam or fire train; cost of Central Sugar Works fitted up completely for making sugar and sirup and working to sugar; semi-sirup made by small sirup outfits, will be willingly made out on application, if the information is desired; also, cost of manufacturing sugar, &c., based only from my own practical experience.

Respectfully yours,

A. J. RUSSELL,
Janesville, Wis.

B. SILLIMAN, Esq., *New Haven, Conn.*

13.—*CRYSTAL LAKE REFINERY.*

CHICAGO, *April* 10, 1882.

DEAR SIR: Seeing the invitation in the Rural World to those interested in the sorghum sugar-cane, and thinking perhaps my experience may be of interest to you, I herewith give you a detailed account of my doings of the past three years.

In the first place let me state to you I am a practical sugar-refiner; spent some eight years in the West Indies making sugar from cane. So you will perceive I came here well armed in the knowledge of the business of sugar-making. In August, 1879, I saw sorghum for the first time, and although the works were put up by inexperienced persons. besides being so near the time for grinding the cane, we had not much chance to make the necessary alterations, so had to get along as well as we could; and as the cane was new to me, and I had little or no faith in its sugar-producing qualities, I resolved to treat it with as much delicacy as a mother would her sick child.

I used only lime for defecating, and made good defecations, equal to the juice from southern cane. I had not polarized the juice, but the defecated juice, while running from the defecators, looked and smelled so natural that I was convinced there was sugar in it: then, reducing it to about 38° B., was obliged to keep it in tanks nearly one month before our vacuum-pan was ready, when I reduced it to sugar. I polarized this sirup, which showed 53 per cent., and am satisfied a good deal of the sugar it contained when fresh had become inverted, as the sirup was slightly acid when we boiled it down to sugar.

Notwithstanding it was boiled at a temperature of 180° Fahr. in the vacuum-pan, in consequence of a short supply of water for condensation. I could readily grain it in the pan. *That* fact alone should decide the question whether sorghum will successfully produce sugar. After the sugar had remained in the crystallizing tanks 48 hours it was so hard that a man weighing 180 pounds had great difficulty in pushing a spade through it. General Le Duc, Malcolm McDowell, and others can testify to that fact, as they were eye-witnesses.

In consequence of the vacuum-pan boiling the sugar so hot, and not being familiar with the juice, and wishing to get as large a yield of sugar as possible, I boiled it rather stiff, which made the grain finer than I wished it; but to the experienced *that* did not detract one iota from its strength. I continued to run till I had made over 50,000 pounds of sugar. In appearance it was good C sugar. It was tested in Boston and New York, and showed 96¼ per cent. of sugar. The above was accomplished in the fall of 1879. We generally designate sugar machinery on a plantation as machinery, but I fail to find a suitable name for the "plant" I had.

In 1880 we had made alterations, in order to do some pretty good work, planted about 300 acres of cane, and a month before it matured it was struck by a hurricane and damaged to such an extent that we received only the product of 30 acres. That mixed with dead cane, rendering the juice so bad, that the sirup only polarized about 42 per cent.;

boiled some for sugar, but finding it very gummy, abandoned the idea, and made only sirup. Thus ends the chapter for 1880. In 1881 the spring was so backward our cane hardly matured, and the sirup from it polarized about the same as the previous year (42½ per cent.). Having such bad luck the past two years at Crystal Lake, Ill., where the above experiments were tried at the works of F. A. Waidner & Co., we have concluded to abandon any further work at the above place. I should here state that Crystal Lake is the most elevated section in the State of Illinois, which makes raising a crop there rather uncertain; although the old residents of the place say they never experienced two such years with sorghum as 1880 and 1881; indeed that is the general verdict throughout the country. Crystal Lake is situated about 44 miles north of Chicago. I am interested in a large works at Hoopeston, Ill., which is attached to a corn canning establishment erected for the purpose of utilizing the cornstalks. *That* we found was *no go*, as the stalks had but little juice; could not produce enough sirup to pay expenses. I consider the cornstalks had a thorough test. We found only about a foot or a foot and a half of the stalk to contain juice; the rest was a dry pith. At the time, the corn was in the roasting ear. The cornstalks were tested in 1880. In 1881 we cultivated 500 acres of sorgo, and the drought was so severe we only got about 2¾ tons to the acre, instead of from 10 to 20. Cane was very thin, and in some instances not over 2 or 3 feet long; sirup only polarizing 40; did not attempt to make sugar. This year we are putting under cultivation at Hoopeston 1,000 acres. We sold all of our product last year, by the car-load, in this city at 50 cents per gallon.

Notwithstanding I have been here three seasons, I have not had a single day's fair trial of sorgo juice. With the plant of machinery we have at Hoopeston now to work up juice such as I had in 1879, I am sure the results I could produce would astonish the country.

I am satisfied of one thing, that the cultivation of the cane is not thoroughly understood. One great drawback here has been the want of proper machinery and a knowledge how to treat the juice. They imagine all that is necessary is to boil out the water and let nature do the rest.

I have been a very careful student for the last three years; and consider myself now familiar with the juice, and just want one fair chance. They were thirteen years in Louisiana before they could successfully make sugar from the Ribbon cane. We did it here in six weeks. If there are any questions I can answer for you I shall be delighted to do so.

Please acknowledge the receipt of this letter and oblige

JOHN B. THOMS.

B. SILLIMAN, Esq., *New Haven*, *Conn.*

CRYSTAL LAKE REFINERY,
Nos. 231 AND 233 SOUTH WATER STREET,
Chicago, April 10, 1882.

DEAR SIR: I had just finished the inclosed effusion when your letter of the 7th came to hand, so will send it as it is written, adding the further information you require.

The juice in 1879 weighed about 8½° Baumé; did not polarize the juice. From a gallon of sirup, weighing 11 pounds, we got a yield of about 4¼ pounds of C sugar from the gallon, and about 46 per cent. of a gallon of sirup weighing 11½ pounds (to the gallon). As the cane crusher was a very miserable affair, we could not squeeze the cane enough, and as no account was kept of the yield of juice, cannot give the percentage. In other words, the works were so miserably arranged it was almost impossible to do anything with system. After I was in a position to keep an account of the yield of sirup per ton I have received as high as 23 gallons and as low as 15 gallons of sirup, weighing 11½ pounds to the gallon, from a ton of cane. The difference in the yield was occasioned by the different densities of the juice. The sirup that produced the 4¼ pounds of sugar to the gallon polarized 53. I have never worked any other cane but the Early Amber. An acre of land has produced, to my knowledge, as high as 21 tons of stripped cane; presume a fair average yield would be 12 tons to the acre. The way we work at Hoopeston is as follows: We lease the land at $4 per acre, and pay a farmer $8 per acre for cultivating and delivering cane to mill unstripped. That seems to satisfy the farmer, and gives us cane at a low figure. We also will pay $2 per ton for cane delivered at mill. I do not strip the cane, but think it would improve it to do so; I think an average crop of cane will pay a farmer better than raising corn.

I would not like to state at what stage a maximum of sucrose may be found, but would advance the idea, just as it is growing ripe. The cane, in some districts, deteriorates very rapidly after being cut. I have kept cane for nearly a month, during cold weather, in Kansas, which did not deteriorate any more than cane cut three days at

Crystal Lake. Cane should be worked up as soon after being cut as possible. The juice spoils more rapidly if the cane is allowed to stand in the ground with the seed top cut off. The juice defecates about the same as Southern cane juice, but in clarifying it produces more scum, and, as compared with Southern juice, it contains more foreign matter. I have used sulphurous acid, which is good in this way: It enables us to use more lime in the defecation, but could produce a better yield of sugar without it, as I have given it a thorough test. Sirup made with sulphurous acid turns sour much' quicker than without. I prefer the sulphur fumes; when I use the sulphur fumes I use lime enough only to neutralize the acid in the juice before receiving the sulphur fumes. We cannot make a thorough defecation with lime, and reduce it to sirup density, without making very dark sirup, unless we use sulphurous acid or sulphur fumes. With the use of lime I eliminate the sorghum tang and make the sirup a perfectly neutral sweet.

You will please understand that there is a difference between a clarified sirup and a defecated sirup—the latter is made by using a defecating agent, while the former is simply accomplished by boiling and skimming. Of course the clarified sirup, such as the farmers make, would not yield as much sugar as the juice properly handled, with the use of lime. I have not yet met a single person here in the West who has the slightest idea how to treat the sorgo juice for sugar. I find the juice more delicate than the Southern cane juice to handle. The way I work is as follows: As the juice comes from the mill it is passed through sulphur fumes, run into tanks, of which there are four, holding about 600 gallons each. After having settled, the juice is run into the clarifier, where it is limed, defecated, and clarified. This clarifier holds the contents of a tank, so that when it is charged it empties the tank. After the juice has been clarified it is run into the evaporator, where the cleansing is finished, boiled down to about 20 F., run into settling tanks, thence to the vacuum-pan, where it is reduced to sugar or sirup; from thence, if boiled for sugar, is run into crystallizing tanks, then purged of its sirup in the centrifugals. To work 100 tons of cane per twenty-four hours, the expense will be as follows (I should here remark that the cost to produce sugar will be no greater than simply to make sirup, for should you get 4 pounds of sugar from a gallon of sirup, you have but 50 per cent. of sirup left, and the difference in the cost of the sirup packages and the sugar barrels will more than pay for the extra labor; I will add coal to the expense, the use of which, in some instances, is superfluous, for the bagasse, or refuse cane, is ample for fuel; for after the bagasse is spread in the sun, one or two days' drying makes it an excellent fuel):

100 tons cane, at $1.50 per ton	$150 00
12 men, feeding mill, at $1.50 per day	18 00
2 men, engineer and assistant, at $4.50 per day	9 00
4 men, firemen, at $1.50 per day	6 00
4 men, hauling bagasse, at $2.50 per day	10 00
2 men, clarifiers, at $1.50 per day	3 00
2 men, evaporators, at $1.50 per day	3 00
2 men, juice tanks, at $1.25 per day	2 50
1 man, fill sirup barrels, at $1.25 per day	1 25
5 tons coal, at $3.50 per ton	17 50
30 sirup barrels, at $1.40 each	42 00
Incidentals	10 00
Total expense to work 100 tons cane for 24 hours	272 25

You will perceive from the above calculation that the sirup will cost 18.15 cents per gallon, taking 15 gallons' yield to the ton. Should the yield be larger, it will reduce the cost accordingly. Say the yield be 18 gallons to the ton, you add the extra cost of packages, and it reduces the cost of sirup to 15.59 cents per gallon, and as our sirup readily sold at 50 cents a gallon, in 300-barrel lots, you will perceive the margin is large enough to enable us to pay more for our cane, if needs be.

To ignore sugar entirely, I know of no business that pays as well, and I know of no better place to work this business than Texas or Kansas, prefer the former, as the product could be shipped directly to New York, Philadelphia, or Baltimore at a less cost than the charge of transporting 300 miles from Kansas by railroad.

Hoping the voluminous appearance of this letter will not frighten you, if there is anything I have omitted which you desire to know, command me.

Respectfully,

JOHN B. THOMS.

B. SILLIMAN, Esq., New Haven, Conn.

14.—*LETTER OF GEORGE W. CHAPMAN, STERLING, KANS.*

STERLING, PRICE CO., KANSAS, *Feb.* 6, 1882.

DEAR SIR: Will you please send me your report of 1881? As sugar-cane is very prolific here, I imagine it would be of vast importance to us, and would like to have the following seeds: The Red Brazilian artichoke, Chinese, White Liberian, Mammoth, and any other good sugar-cane seed. I am also anxious to try some cotton and Beauty of Hebron potato.

I worked up last season 75 acres of cane, Amber and Honduras. Amber yielded 9 tons stripped and topped, and the Honduras 33½ tons raw stalk per acre, being the largest yield of cane yet known in Kansas. I can substantiate the above yield by the affidavits of four responsible men, and in this vicinity no one disputes it who has seen it. If a full report of cultivation, &c., will be of any value to you, I will forward the same. I made some sirup by an evaporator and it all granulated in a solid.

Can you give us some information where we might prevail in getting some capitalists to come and develop scientifically the manufacture of sugar?

Kansas can grow the deficit if it can be manufactured.

If you have any improved methods or seed of any kind, we will feel highly pleased to receive it.

Your obedient servant,

GEO. W. CHAPMAN,
Secretary Price County Farmers' Club.

P. O. box 170.
COMMISSIONER OF AGRICULTURE.

15.—*LETTER OF JOEL M. CLARK, ITALY HOLLOW, N. Y.*

ITALY HOLLOW, N. Y., *March* 8, 1882.

SIR: The Amber sugar-cane seed that I received from the Department of Agriculture on the 17th day or April, 1881, I planted on the 13th day of May last upon dark gravelly soil—planted in drills 3½ feet apart; used no fertilizer except ashes, which I used at the rate of about 12 to 15 bushels per acre. Cultivated and hoed twice, same as corn. At second hoeing pulled off the suckers, and on the 14th day of September I commenced cutting the cane for sirup, which was from 9 to 11 feet high; the heads were very dark color, the seed hard and appeared to be fully ripe. The yield was 13½ tons cane per acre, from which I made 18½ gallons of beautiful sirup per ton, pronounced by dealers equal to the best sirup in the market. From the sirup I obtained a large per cent. of crystallized sugar, which was very satisfactory. I also obtained about 20 bushels of seed per acre.

Whole experiment entirely successful. There will be from 40 to 60 acres planted in this town the coming season.

Respectfully yours,

JOEL M. CLARK.

COMMISSIONER OF AGRICULTURE, *Washington.*

16.—*JOSEPH WHARTON; HIS POOR SUCCESS WITH BEETS FOR SUGAR; SORGHUM PROMISES BETTER. (LETTER TO THE CHAIRMAN.)*

AMERICAN NICKEL WORKS,
Camden, N. J., April 8, 1882.

DEAR SIR: Your inquiry of 4th instant as to results attained in my (beet) sugar experiments at Balsto is received. Those results were but negative.

The first year I selected a level piece of sandy ground, similar to most South Jersey land, and dividing it into a number of parallel strips manured them with various fertilizers, namely, barn-yard manure, green sand marl, kainite, fish scrap, and swamp muck. The beet seeds were planted, at different dates, across the other strips.

Cultivation was carefully attended to; but the crop was very light, though part of it was about 8 per cent. in sugar. No results could be deduced from the variety of fertilizers, the crop being generally poor.

The second year I employed Mr. Gaston Barbier, who had had charge of a beet farm in

France, to manage the affair from early spring, viz, the preparation of ground, until the crop was gathered. The seeds I bought direct from a famous beet-seed cultivator in Saxony and every detail was left without restriction of expense to Mr. Barbier, who confidently expected a good yield of beets and of sugar. Again the crop failed, and more signally than before. After that, instead of putting in 10 of 15 acres, my farm manager has planted but a few in the garden, and has merely raised a few beets for the table.

My judgment is that the soil of South Jersey, or at least that part of it where sand is underlaid with gravel, is really too poor to carry a good crop of beets, even when well manured; also, that the climate of that region is unsuitable, because a dry spell usually comes on after the plants have fairly set and begun to grow vigorously; this checks and stunts the growth, prevents the leaves from spreading over the ground in such way as to serve as mulching, and the continued drought has full opportunity to dry out the light soil. Where the beets survive this trial and attain a certain magnitude, they take a second growth upon being thoroughly wetted by later rains, and that destroys the normal texture of the beet, causing it to be fibrous, watery, and low in saccharine.

Cattle fed upon either the beets or upon the "pugs," that is, the roots after rasping upon a French machine and draining in a centrifugal, did not thrive or fatten very well; they inclined to scour and were not solid in flesh.

Sorghum promises much better and I have some faith in the possibility of South Jersey to produce sugar from it to advantage; the plant grows well with just such soil and treatment as maize.

Yours truly,

JOSEPH WHARTON.

Prof. B. SILLIMAN.

17.—LETTER OF J. F. PORTER.

RED WING, MINN., *November* 5, 1881.

DEAR SIR: Yours of November 1 received. The season of 1880 was my first experience in the sugar business. That year I made about 4,000 pounds; sold it for 9 cents per pound. This year I have made some, but a very small amount. This was made in the forepart of the season before bad weather set in; after that the percentage of sugar was so small I abandoned the idea of making sugar. I have made this season 10,300 gallons of sirup, most of which is an excellent quality and is selling for 45 and 50 cents per gallon by the barrel. Don't know of any one else that has made any sugar in 1880 and 1881.

Yours truly,

J. F. PORTER.

Mr. PETER COLLIER,
    Washington, D. C.

18.—LETTER OF BLYMEYR MANUFACTURING COMPANY.

BLYMYER MANUFACTURING COMPANY,
    Cincinnati, March 15, 1882.

DEAR SIR: Your circular noted. We mail our sugar H. B.; also circular of machinery.

That sugar can be made from several varieties of sorghum is of course established. We have known of its being made by the barrel as far back as 1862. We have known several farmers who have made one or more barrels of it several consecutive years. It remains to be seen whether sorghum can be depended upon for sugar as it is for good sirup.

The enormous value of the sorghum crop even in sirup would, we suppose, astonish the country if reliable statistics could be had. There is certainly warrant enough in what has already been done in sugar-making to justify special scientific investigation.

Yours truly,

BLYMYER MANUFACTURING COMPANY.

Mr. B. SILLIMAN,
    New Haven, Conn.

19.—LETTER OF H. W. WILEY.

THE LA FAYETTE SUGAR REFINERY,
La Fayette, Ind., April 3, 1882.

DEAR SIR: I would suggest that in the investigations of your committee especial attention be given to the influence of sulphurous acid on the color, flavor, and quality of the product. Also the solubility of the lime compounds formed by the neutralization of the acids of the juice, and the possibility of removing them wholly from the finished products.

If your final report could be delayed until next winter it would give an excellent opportunity for the investigation of some of these unsolved problems during the coming summer and fall. My own investigations in the directions indicated are still too imperfect to lay before your committee, but I hope by another year to have them in a much more complete shape.

All persons interested in sorghum culture will take a lively interest in your labors.

Respectfully,

H. W. WILEY.

Prof. BENJ. SILLIMAN, *Chemist, &c.,*
    New Haven, Conn.

20.—LETTER OF JOSEPH ALBRECHT, CHEMIST.

[From the New Iberia "Sugar Bowl" of September 15, 1881.]

SORGHUM SUGAR.

REMARKS ON THE REPORT OF PROF. P. COLLIER, CHEMIST OF THE AGRICULTURAL DEPARTMENT AT WASHINGTON.

I was always of opinion that no plant could compete with sugar-cane, and that any attempt to manufacture sugar either from beet-root or sorghum would end in financial failure in this country. My opinion was much modified when three years ago I had occasion to experiment on Amber cane. The plants were in bloom, still the juice of the crushed stalks had a density of $12\frac{1}{2}°$ Baumé and contained 10 per cent. of prismatic sugar and 5 per cent. of glucose. I was much astonished at the unexpected richness in sugar in that variety of sorghum.

In a recent visit to the Agricultural Department in Washington I had the good fortune to become acquainted with Prof. Peter Collier, who kindly introduced me into his laboratory. There he pointed out to me many samples of sugar made from as many varieties of sorghum and of cornstalks; described his mode of expressing and clarifying the juice, of boiling it into sugar; he also explained to me his method of analyzing for saccharose and glucose, which he pursued from the earliest apparition of the stalk until after ripening of the seed. He showed me his voluminous manuscript, proving by numberless carefully executed analyses that certain varieties of sorghum can develop as much sugar as the true sugar-cane of the South, and that he succeeded in determining by continued analyses the state in which the juice has reached its maximum richness in saccharose. The astonishing results which the professor obtained in his experiments convinced me that sorghum will be the future plant from which the Middle and Northern States will make their sugar and molasses, and that the "dream" of Commissioner Le Duc, to save the country many millions of dollars which now go to foreign countries, *will be ultimately realized.*

I am now in receipt of the printed "report of analytical and other work done on sorghum and cornstalks by the chemical division of the Department of Agriculture under direction of Commissioner W. G. Le Duc, by Peter Collier, chemist." This report is well worth more than a fugitive perusal; it contains matters of the greatest interest, not only to the agricultural and commercial world, but also to the political economist, I may say to every man, woman, and child of our country. Professor Collier undertook a work which, for its magnitude, thoroughness, and clearness in matters pertaining to the cultivation of sorghum, for the purpose of manufacturing sugar, stands alone and far ahead of anything which has been done by the Agricultural Department.

Commissioner Le Duc could not have confided this part of his "dream" to a more able and enthusiastic man than Prof. P. Collier. Never before has a plant been so carefully studied in all its phases in the development of saccharine matter, together with the scientific determination of other substances besides saccharose which interfere more or less in the manufacture of sugar.

The amount of labor, patience, and scientific skill bestowed on this work can scarcely be estimated by the average readers, but suffice it to say that the author solved the problem of making sugar economically from certain spécies of sorghum, and fixed the period of growth at which the stalks are the ripest for the mill. With this report in hand, the planter and sugar-maker have a sure guide of the growth and gradual development of sugar in the cane, and need no more waste their time and money in experimental groping in the dark.

The cultivation of the richest variety of sorghum will henceforth become more and more extended, until it will supply the wants of the whole country. With the help of the description and the thirteen plates of very neatly executed wood-cuts with which the report is adorned it will be easy to determine the *variety* of sorghum experimented on: but the most important part of this work are the careful analyses of the laboratory, demonstrating the period at which the juice of each particular variety of sorghum or corn contains the most crystallizable sugar which could be profitably separated.

Not less valuable are the synoptical tables of the varieties of sorghum cultivated at the Department of Agriculture, showing the average composition in each stage of their growth, and the graphical plates, revealing at a glance the whole history of the changes of the juice in the different varieties; the development of saccharose, glucose, and other solid matter is made visible at once as the plant progresses towards maturity, rendering it henceforward easy for the cultivator to judge of the proper time to cut his cane for the mill. We know, from our own experience, that the greater the specific gravity of the juice the richer it is in saccharose and the less in glucose and other foreign matter.

Professor Collier has made the same observation, and has fixed, as a rule, that when the juice has reached the density of 1.066 specific gravity (equal to 9° B.) the stalks may be considered ripe enough to be cut for grinding.

The report contains also a table of comparison of the different hydrometers, which will be much appreciated by those who possess not the instruments or other means to ascertain otherwise the specific gravity.

Considering the cheapness of the seed of the sorghum, the facility of its cultivation, the rapidity of its growth, the wide range of climate in which it can be successfully raised, and the abundance of seed, with many other advantages over the southern sugar-cane, it must engage our serious meditation, and I believe that the prediction that sugar and molasses will soon be manufactured from sorghum throughout the United States is not too much ventured.

JOSEPH ALBRECHT,
Chemist, 14 Union Street, New Orleans.

21.—LETTER FROM MR. RANSOM TO THE CHAIRMAN, IN RESPONSE TO AN INQUIRY, OF DATE OCTOBER 22, 1882.*

SALEM, RICHARDSON COUNTY, NEBRASKA, *October* 22, 1882.

DEAR SIR: Your memorandum is received.

Balance-sheet for 1881 should foot up as follows to be correct:

| 14 acres Early Orange: | { Credits | $786 20 |
| | { Expenses | 352 50 |

| Profits | 433 70 |

That is as it should appear in your appendix—as my results from 14 acres for the year 1881—a season so dry in this section that wheat alongside of it and on just as good ground made 2 bushels per acre, and corn 10 bushels, the cane leaving me a net profit of $30 97 per acre, while the wheat and corn made a heavy balance in the wrong column on my ledger.

Now for this fall's work. I am not through yet and can only give you partial results.

Six acres of Amber cane made 183 gallons per acre, which left me a profit of $63.70 per acre.

I have 26 acres Early Orange, of which about 6 remain to be worked yet. I think it will average about the same, some of it making more per acre and some considerably less. A complete balance-sheet will probably change the profits a little on the above figures for the 6 acres of Amber.

---

* This letter of Mr. Ransom refers to one of earlier date, which follows, and corrects a clerical error in the former statement.

I have $1,500 invested in the business (factory). My juice weighs from 10° to 12° B. I use 3 gills of milk of lime in the juice tank when about one-fourth full, and the same when the same juice is run into the defecator. That is for 150 gallons of juice. In the defecator it is brought to 208° Fahr.; then well skimmed; then boiled gently to raise the heavy brown scum that is taken off. Then the juice is run into settling tanks, where one pint of burnt alum water is added. From settling tanks it is drawn from the top through swing pipes into the evaporator. I use an open fire train. I find a ready sale for my sirup at home at 50 cents per gallon at wholesale, and pay for the barrel. It is granulating now. Some made yesterday (boiled to 220°), was so heavy at 9 p. m. that the bottom of the day's run would not go through the perforated tin strainer that we use in straining through into a 2,600 gallon reservoir. I never had any experience in the business until two years ago, when I made a failure in my first year's operation. My second year's results I give in the first part of this letter, and, if not too late to do you any good, I should be pleased to send you a copy of the balance-sheet for 1882 as soon as I get through.

Any information you might be able to give me that would help me to separate the sugar from the sirup this fall would be a personal favor, as I am satisfied that I could take out several thousand pounds from my reservoir if I knew just how to go about it.

Very respectfully, yours,

B. V. RANSOM.

Prof. B. SILLIMAN,
    New Haven, Conn.

22.—LETTER FROM B. V. RANSOM, SALEM, NEB.

[From the Rural World.]

A SORGO BALANCE-SHEET AND LARGER MILLS.

Colonel COLMAN: As Mr. Day has sent you a partial copy of my 1881 balance-sheet for publication, perhaps it would be as well to send you the sheet complete, as it will give your readers a better understanding of the profits and capital invested from sorgo-growing than they can get from that:

By sirup sale		$744 20
By 140 bushels seed, at 30 cents		42 00
		786 20
Man and team listing 14 acres, 2½ days, at $2	$5 00	
Man and horse drilling seed, 2 days, at $1.50	3 00	
Man and team harrowing, 3 days, at $2	6 00	
Man and team cultivating, 12 days, at $2	24 00	
14 cords wood, at $4	56 00	
Chemicals, oil, and blacksmith	6 00	
Rent of 14 acres grain, at $3	42 00	
Cost paid for labor to manufacture	131 00	
Board of hands	65 00	
		352 50
Net profit from 14 acres		433 70
Net profit per acre		30 98
Net expenses per acre		25 17
Total yield per acre		56 12
Net wholesale price per gallon, sirup		50
Cost per gallon to manufacture		23
Net profit per gallon		27
Capital invested		1,000 00
Per cent. interest on capital		43

The figures above are actual results from working on a small scale. My mill was entirely too small for the balance of my work. If the capacity of that had been equal, the profits on capital invested would have been more than double what they were.

I used a two-horse mill that weighed about 1,300 pounds. It crushed about half it was recommended to per hour, and what is worse it had one weak spot, that is in the knife or return plate; and as the strength of any mill is the strength of its weakest place, my mill only worked about half what was claimed for it.

My cane juice weighed 12° B. during the entire working season. Five gallons of juice made 1 of sirup that when cold marked 42 B., yet I got only 12 gallons of sirup per ton of cane. Will Kenny got 12¾ gallons per ton from cane that the best of it marked only 9° B., and run down as low as 3, and Schevrey & Beecher 16 gallons from cane that averaged 10° B.

The difference is they use a stronger mill and probably get 50 or 70 per cent. of the juice of the cane crushed, while I got only from 30 to 40 per cent.

B. V. RANSOM.

23.—LETTER FROM EPHRAIM LINK.

[Author of "Link's Hybrid."]

MYSTERIES OF SORGHUM.

Colonel COLMAN: Reading A. S. Folger's article in Rural World of December 29, reminds me of what I have for many years thought, viz, that there are as many mysteries and hidden capabilities in the sorghum family as in any production of the soil. My first convictions on this subject grew out of an experiment I made a good many years since, in an attempt to make vinegar out of the juice boiled barely to a clarifying point. The result was a mystery, which, however, I did not set out to detail in this article, although not without its points in the curious chemistries of the sorghum family. But as I may have furnished the seed the Department sent to Mr. Folger, I wish to say that perhaps six years since I procured my first Honduras seed from Mississippi, and readily found it much superior to any of the varieties I had before cultivated, and discarded all others in the endeavor to prevent any hybridization. It remained seemingly pure and fully satisfactory for several years, during which time I furnished the Department at Washington seed for distribution to the amount, in three years, of 50 bushels or more. In my crop of 1879 I saw a good many heads indicating a mixture, for which I could not account, and which I had been so careful to avoid, unless the contamination occurred the first year, when another variety grew a little distance off. If so, the contaminating principle lay dormant three years and had developed only that year. I sent to a friend in Texas for an entire renewal of seed for the planting of the spring of 1880, and found that, and the crop of last year, to be very pure, and to ripen two or three weeks sooner than the same variety before grown. Here also is a locked mystery I fail to understand. Also, four years ago I found a head—a clear sprout in the Honduras, entirely different in appearance from it, propagated it, and found its yield and richness in juice second to no other, and its sirup freer from the sorghum flavor than any I ever made. I sent General Le Duc a specimen of the sirup and seed, and he ordered all the seed I had, about 14 bushels. In his report of the analysis of varieties he calls it "Link's Hybrid." It grows to good size, stands well, ripens before the Honduras, and I predict for it a high place among varieties.

EPHRAIM LINK.

GREENVILLE, TENN. *

24.—LETTER OF ISAAC A. HEDGES, SAINT LOUIS, MO.

This letter, of date April 12, 1882, to B. Silliman, chairman, accompanying a considerable collection of samples of sugar, melada, and sirups, being the same which were exhibited at the Cane Growers' Association in the previous January at Saint Louis. Mr. Hedges's letter also contains valuable data, to which reference is invited.

* It will be observed that Mr. Conrad Johnson, in his valuable letter annexed (XVI), makes special mention of Dr. Collier's analysis of this variety, the origin of which Mr. Link here explains.

B. SILLIMAN, Chairman, &c.

25.—*LIST OF SAMPLES EXHIBITED BY THE CHAIRMAN OF THE COMMITTEE.*

[Deposited in National Museum April 19, 1882.]

(A.) FROM DEPARTMENT OF AGRICULTURE.

Sorghum sugar, Department of Agriculture, 1881, 97°.5 polarization.
Sorghum sugar, Department of Agriculture, 1881, 92°.6 polarization.
Sorghum sugar, Department of Agriculture, 1881, 86°.4 polarization.
Cornstalk sugar, Department of Agriculture, 1879, 81°.6 polarization.
Pearl millet sugar, Department of Agriculture, 1878, 73°.4 polarization.
Sorghum seed, White Mammoth.
Sorghum seed, Early Amber.
Sorghum sugar, Professor Swenson, 1881, 96°.4 polarization.
Sorghum sugar, A. J. Russell, 1880, 97° polarization.
Sorghum sugar, Hilgert & Sons, N. J., 3 samples.
Sorghum sugar, Faribault Refinery, R. Blakeley, Minnesota.

(B.) FROM THE MISSISSIPPI VALLEY CANE GROWERS' ASSOCIATION, BY MR. ISAAC
A. HEDGES.

Sorghum sugar, A. Folger, Washington, Iowa, 3 samples.
Sorghum sugar, Clinton Bozarth, Cedar Falls, Iowa.
Sorghum sugar, Bartlett, 1880.
Sorghum sugar, Thorp, New Haven, Conn., 1880.
Sorghum melada, A. Folger, Washington, Iowa.
Sorghum melada, Captain Brown, West Baton Rouge, La.
Sorghum melada, E. W. Deming, Byron, Ill.
Sorghum melada, S. M. Poland, Sandusky, Iowa.
Sorghum melada, William Frazier, Enterprise, Wis.
Sorghum melada, No. 3, Virden, Ill.
Sorghum melada, Baton Rouge, La.
Sorghum melada, L. M. Thayer, Kenosha, Wis.
Sorghum melada, Thorp, New Haven, Conn, 1880.
Sorghum melada, Bartlett, North Guilford, 1880.
Sorghum melada, Clinton Bozarth, Cedar Falls, Iowa.
Sorghum melada, Jacob Stine, New Madrid, Mo.
Sorghum melada, Charles Ranch, Virden, Ill.
Sorghum melada, W. D. Clark, Colfax, Ill.
Sorghum melada, J. M. Nash, Hudson, Wis.
Sorghum melada, N. A. Layton, Giles Mills, N. C.
Sorghum sirup, L. M. Thayer, Kenosha, Wis.
Sorghum sirup (B), J. A. Sebold & Co., Great Bend, Kans.
Sorghum sirup, J. A. Sebold & Co., Great Bend, Kans.
Sorghum sirup (A), Oak Hill Refinery, Edwardsville, Ill.
Sorghum sirup, Jesse Allen, Washington, Ohio.
Sorghum sirup, A. Folger, Washington, Iowa.
Sorghum sirup, N. A. Layton, Giles Mills, N. C., 2 samples.
Sorghum sirup, F. Kingsley, Hebron, Nebr., 2 samples.
Sorghum sirup, Port Huron, Mich.
Sorghum sirup, unknown source.
Sorghum sirup, Anthony, Kans.
Sorghum sirup, Ovid, Mich.
Sorghum sirup, Seth H. Kenney, Morristown, Minn.

26.—*LETTER FROM THE LATE ISAAC A. HEDGES, EX-PRESIDENT MISSISSIPPI VALLEY CANE GROWERS' ASSOCIATION.*

SAINT LOUIS, Mo., *April* 12, 1882.

DEAR SIR: I have the honor to acknowledge your favor of the 1st instant, inclosing circular of sorghum sugar, &c.

It affords me great pleasure to learn that your enlightened academy have taken the investigation of this industry in charge.   In the language of our able Commissioner, Dr.

Loring, "I doubt not much good will come of the investigation." If the products of this crop cannot receive your indorsement, after the most rigid scrutiny, then the people sho uld know it; but if, however, it should (as I trust it will) come out of the scientific crucible with that measure of commendation that it has received at the hands of many of our State professors, as well as the commercial public, then capitalists will embark in it, and skilled operators will be employed in the business. The results will be the production of an abundance like the best samples I have the pleasure of sending for your inspection.

The several samples I forward to you by express are, with three exceptions, those that were delivered to our association at its late meeting by the members who produced them. I prefer to forward them mostly in their original packages (especially the sirups) as affording you a better evidence of their original quality and genuineness. I have labeled some of them with letters and corresponding ones in the explanatory table. I am compelled to do so because of the lack of any record or printed description to furnish you. The four meetings of our association have been, with few exceptions, of new beginners, and hence with limited reports to make. This will account for our lack of statistics, or recorded essays. I send you a copy of the report of our last proceedings, which, though limited, may possess some merit.

I send you a copy of my third revised edition of "Sugar Canes and their Products, Culture, and Manufacture."

I call your especial attention to the reports of J. S. Lovering, sugar refiner of Philadelphia. Pa., as copied in the body of my work, pages 123 to 140. I had the satisfaction of examining his samples of sugar and molasses at the time, and can testify to their being equal to any of our refined sugars of to-day.

I will also call your attention to the report of Mr. Charles Belcher, of the Belcher Sugar Refining Company of Saint Louis, on page 173, chapter 14, together with Mr. Thom's (a practical sugar boiler) discussion of Mr. Belcher's report (pp. 173, 174).

SUGAR SAMPLES.

Among sugar samples, I will only call your attention to a few of them.

First. That of Mr. C. Bozarth, of Cedar Falls, Iowa, whose report will be found in a copy of the proceedings of our late cane growers' meeting, page 19, to which I will invite your attention. His simple method and uniform success give to the student of this industry greater assurances of ultimate success than the same would from the hand of a scientific expert.

The crystals formed in the Orange cane, so called, will generally be found much larger and more cubical in form than the Amber. This latter variety of cane as well as the Orange are outgrowths by hybridization of some of the imphees imported into this country by Leonard Wray, a sugar master and author then of London, England, in the year 1857. It will be observed that all of the samples I send you, with one exception, are from these two varieties, both of which are the result of mere accidental culture. Their superiority is so manifest that they are rapidly superseding all others.

It is fair to conclude that as soon as the well-directed efforts of the skilled cultivators of our State and national agricultural schools are applied to the further development of these several varieties of canes by hybridizing, and the selection of seed from precocious and well-developed canes only (something which has not yet been practiced, to my knowledge), then the most satisfactory results must invariably follow.

SIRUP SAMPLES.

My principal object in placing in your possession the many samples of sirup is that your honorable committee may learn how varied the results are from the same product in the hands of so many different operators. This varied quality (the result of want of skill and good apparatus) has prevented this sirup from becoming a commercial article. (See communications below on this subject.)

It will be observed that in all cases where the samples approach a straw color and have been properly treated with an alkali to neutralize the free acid, granulation in a greater or less degree has resulted.

I will call your attention to the large package marked "two years old." This formed no crystals the first season, and now they are peculiar, though not abundant. They are characteristic of the Orange cane.

The introduction of the sulphur fumes in the juice of the sorghum has been tried to some extent the past season. I send you three samples, marked A, B, C, and respectfully ask your investigation of the sanitary qualities of each. Sulphur fumes are extensively used in Louisiana, and may be beneficial in the sorghum; but there may be danger of an imperfect reaction of the acid, by which a portion is retained in the sirup,

S. Mis. 51——9

that may be injurious to health.   One of the above samples, I think, indicates that condition.   Hence I am more particular to call your attention to it.

The proneness of our people to engage precipitously in any new enterprise suggests the importance of a censorship somewhere, or rather an umpire, to which doubtful questions may be submitted at all times, and particularly in those matters so deeply affecting the general health and prosperity of the whole nation, as does this industry.

The following are the communications referred to above, all of which are respectfully submitted:

[Office of Brookmire & Rankin, wholesale grocers, 415, 417, and 419 North Second street and 213 and 215 Locust street.]

SAINT LOUIS, *April* 10, 1882.

DEAR SIR: Your favor of the 10th to hand, and we would reply to your questions as follows:

Question 1. Is there a demand in this region for sorghum sirup as now manufactured?

Answer. There is a good demand at all times for sorghum sirup, but the trouble has been the want of care in its manufacture and lack of knowledge, the result being goods uneven in quality, as well as body, and in many cases scorched.   Another great drawback has been bad and dirty cooperage.   We are pleased to state, however, that the goods bought of you have been No. 1 in quality; cooperage new and uniform, and in every way desirable.

Question 2. From what source does the demand arise, if any?

Answer. The demand is not confined to any particular locality.   A good article is wanted from all sections of the country.

Question 3. How does this sorghum sirup, as now manufactured, compare with other sirups and molasses, say "Sugar-House," "New Orleans Plantation," "New Orleans Centrifugal," and glucose sirups of various names and grades?

Answer. The sorghum as you, or rather your friends, make it would class nearer the old process New Orleans molasses.   It does not class or conflict in any way with sugar, corn, or glucose sirups.

Very respectfully,

BROOKMIRE & RANKEN.

Mr. ISAAC A. HEDGES,
        320 *Monroe street, City.*

[Office of Wulfing, Dieckriede & Co., wholesale grocers, Nos. 7, 9, and 11 North Second street.]

SAINT LOUIS, *April* 11, 1882.

DEAR SIR : In answer to the inquiries made in your esteemed letter of the 10th instant concerning the sale of sorghum molasses, we beg to state that we find the article growing in general favor, finding a ready sale where once introduced.

We think the demand for it is for table use principally, as well as for use of bakers, for which purpose it appears as well adapted as the Louisiana Plantation or Centrifugal molasses, which latter it approaches nearer to taste than any of the other kinds of sirups in the market.

In conclusion, would say that we have been well pleased with the sorghum received through you, our experience with it being quite satisfactory from a mercantile standpoint.

Yours, truly,

WULFING, DIECKRIEDE & CO.

Mr. I. A. HEDGES.

[Office of J. F. Weston, baker, 611 and 613 Morgan street.]

SAINT LOUIS, *April* 11, 1882.

DEAR SIR: In answer to your questions, I reply from a baker's standpoint.

1st. The supply of imported sorghum is not equal to the demand.

2d. It must, necessarily, take the place of Plantation molasses, which is becoming more scarce every year.

3d. Improved sorghum, in my judgment, is the only substitute for Plantation molasses. All other sirups, either cane or glucose, lack the necessary acid.

Respectfully,

J. F. WESTON.

ISAAC A. HEDGES, Esq.

I will a dd an extract from a letter just received from Col. II. B. Richards, of Lagrange, Tex., that foreshadows much additional importance to the Orange cane:

" But now let me tell you about my Orange cane. It is no longer doubtful at all but that the Orange cane will become in this climate perennial, and after this year I will only plant every two years. I have tested it now effectually for two years, and am convinced that the stubbles will stand colder weather and more of it than those of the Ribbon cane.

" My cane from last year's stubbles has larger stalks, is taller, and in every way ahead of the earliest seed cane at this time, and I believe will carry its superiority clear through to the sirup barrel. I want to begin grinding by June 1.

" Please let me hear from you soon.

" Yours, truly,

"HENRY B. RICHARDS.

" LA GRANGE, FAYETTE COUNTY, TEXAS, *April* 8, 1882."

A l of which is respectfully submitted.

ISAAC A. HEDGES.

Prof. B. SILLIMAN,
Chairman Com. N. A. S., New Haven, Conn.

27.—LETTER FROM C. CONRAD JOHNSON.

BALTIMORE, MD., *March* 30, 1882.

SIR: In compliance with the request of Prof. Peter Collier, of the Department of Agriculture, Washington, D. C., I have the honor of submitting for your consideration a det iled exposition of my views on the subject of sorghums with reference to the prospective production of sugar from their juices, as appears to me from an examination of his report upon this matter.

Preliminary, however, to any remarks that may be made upon the respective data obtained by Dr. Collier, it must be admitted that his results as therein presented are rathe those of an analytical character than the product of actual experiment, at least to any great extent or beyond the precincts of the laboratory, and that while much valuable scientific knowledge has thus been gained a wide field for examination still remains to be explored before sufficient information shall have been acquired upon which to base a satisfactory verdict in the premises.

Wi hout entering into a discussion of the special methods followed and processes employe l for the purpose of obtaining the various data presented, we may without cavil accept them from Dr. Collier's hands as being scientifically correct. After dealing with this portion of the report, Dr. Collier at once enters upon an interesting and exhaustive examination of the constituent history of the different varieties, and exhibits the results obtained from the analyses of thirty-eight different kinds, made at successive periods during their growth and arranged in tabular form. From these tables and the succeeding ones, where a comparison of the various stages, especially the working periods are shown, and the condensed table (No. 88) of the stages, as determined from the results of the same stage for all the varieties of sorghum, we are enabled to deduce the facts laid down by Dr. Collier as the resultants obtained from them. We thus find that the earlier stages exhibit a minimum amount of crystallizable sugar (sucrose) present in the juice, and that in proportion as the plant advances the amount increases, until it attains a point as high at 12 or 16 per cent. of the juice. The "solids not sugar" likewise are found to increase, but not in a similar ratio to that of the crystallizable sugar.

As regards the glucose, or uncrystallizable sugar, it is to be remarked that during the primary stages we find the amount existing in the juice to be as high as 4.50 per cent. (see ta ble 88, stage 3), while in stage 14, or about the commencement of the working period. it averages only 1.88 per cent., as compared with 11.76 per cent. of sucrose, and in the variety " Hybrid " from E. Link, we find these components in the same period to be 0 82 per cent., or less than 1 per cent. of glucose as against 14.28 per cent. sucrose (table 82), showing a gradual diminution of glucose as the sucrose increases, and as a consequence of these changes a constant increase in the exponential value or purity of the juice. The percentage of juice extracted from the stripped stalks is also given, and is useful, when taken in connection with the table of specific gravities, in the calculation of various data, as also for comparison with similar products from the Louisiana and West I dian cane.

The graphical plates also claim attention, inasmuch as they exhibit more forcibly and clearly the changing composition of the canes during the period of their growth than can be obtained from the tabular statements, as we are at a glance enabled to comprehend the exact composition of the canes at any period within the observed limits, and the relations existing between the sucrose, glucose, and solid matter forming the juices, together with the relative amount of these components when judged by the average proportions of these substances as found in the sugar beet and cane.

Plate XIV of this series presents the average development of the thirty-eight varieties whose histories are given in the previous tables, and is virtually a succinct graphical resumé of the average of these canes during each of the periods into which its growth has been divided, together with an exposition of the "average available sugar per acre, specific gravity of, and purity of juice, and the number of analyses determining the above for each of these stages."

As noticed by Dr. Collier (p. 78, § 3), "it is developed that 'the amount of solids not sugar' increase regularly with and with almost the same rapidity as the glucose diminishes. Thus for specific gravities between 1.030 and 1.086 the average per cent. of glucose is 2.84, and of 'solids not sugar' 2.71, while the actual loss of glucose is 1.76 per cent. and the actual gain of 'solids not sugar' is 2.77 per cent. From the small number of ash determinations (34) it appears that the average per cent. of ash in sorghum juice amounts to 1.07; hence we must conclude that a loss of 2.76 per cent. of glucose is apparently counterbalanced by a gain of 1.70 per cent. of organic solids not sugar, the ash varying but slightly. * * * One point, however, seems to be suggested strongly namely, that the decrease in glucose bears a much closer relationship to the increase of organic 'solids not sugar' than to the increase of crystallizable sugar. In other words, it seems at least possible that the commonly accepted idea that cane sugar is formed in plants only through the intervention of glucose may be a mistaken idea."

In regard to the above remark pointing to the existence of cane sugar as a primary formation in the plant, while much may be advanced in favor of this position, there is on the other hand much valuable evidence, and particularly that of McCulloh, who, after extended examinations into this matter, seems to favor the opposite theory. This authority states in his well-known report to Prof. A. D. Bache on investigations in relation to sugar and its manufacture, February 27, 1847 (p. 188, § 19 and 20), after presenting the results of a series of experiments made with different varieties of cane, the following: "The preceding experiments were instituted with the view of determining whether, as has of late been maintained, cane sugar is a primary or a secondary product in the development of the plant; whether, in other words, it is formed directly by physiological agency from inorganic matter or whether it has existed previously as a constituent and inorganic portion of the cane during its earlier immature condition, and has assumed the nature of crystallizable sugar by virtue of some chemical change at an advanced stage of the development of the plant. The facility with which those vegetable substances undergo transformation, which are composed of carbon, hydrogen, and oxygen, when the two latter are in the proportions requisite to form water, naturally suggests the hypothesis that sugar may be formed in plants by some means analogous to those employed in converting starch and woody fiber into grape sugar and the latter into alcohol and carbonic acid. Yet as all these changes seem but steps of degradation in the scale of organized being, partial returns to organic nature, and as it has hitherto been found impracticable by artificial means to form crystallizable sugar, which seems to occupy a higher rank in the scale of vegetable organization, it has been with seeming probability concluded that cane sugar in the growing plant *results from no chemical reaction* upon such substances as starch, woody fiber, &c., but is produced *directly by some vital* and mysterious forces different from those of mere chemical affinity, and for the discovery of which we should study rather the effects of light, electricity, &c., than mere reactions of acids, alkalies, and other chemical reagents.

The probable conclusion which I would deduce from the preceding analysis, *is that since no other sugar than cane sugar exists in the mature joint of cane grown under favorable circumstances, while left-polarizing sugar is a constituent element of new and immature joints,* which also vary chemically from the former in other respects, and undergo a change not unlike that which takes place in the ripening of fruits, *cane sugar is therefore formed as a secondary product, and probably from the left-polarizing sugar,* an opinion the opposite of that which has been of late generally accepted, and based upon investigations made by M. Hervy in France, under the unfavorable circumstances of using canes grown in hot houses, as well as upon the above-mentioned theoretical reasons. M. Hervy states that there was no difference in the old and new joints of the cane he used, but to every planter it is a familiar fact that new joints differ entirely from old in the countries adapted to the culture of the cane, the former being succulent in fiber and of an astringent taste, destitute entirely of sweetness, while the latter are solid in fiber and contain a perfectly sweet juice."

This explanation of the existence of glucose in the cane, which is further proven by the analyses of Caseca and Icery, who both found traces of this element in their examinations of even mature canes, corresponds with the above opinion. The personal experience of the writer also supports the same conclusions, for the reason that he has found that mature canes on the Island of San Domingo will not at times produce crystallizable sugar, and particularly during those seasons when an extended drought has prevailed, during the latter portion of which the canes have ripened, but which have been followed by rains, causing the canes to take a new growth, and consequently forming a large amount of glucose, whose presence in the juice when expressed and exposed to the rude processes of manufacture generally employed on the island has been sufficient to counteract whatever of crystallizing power there remained in the concentrated sirups. If such be the case, and it seems to be so, we must attribute the increase of "solids not sugar" to some other agency than the one suggested. Moreover that the presence of glucose in the sorghum juices may be reduced very materially, if not altogether, is shown by an examination of the variety known as "Link's Hibrid." (Table No. 12, 1880, to which reference has already been made.)

Table No. 89, distinguished by Dr. Collier as the "specific gravity" table, is, from a practical point of view, one of the most valuable in connection with this subject of any presented to our consideration as being an exposition of the constituencies of sorghum juices, as developed by analysis from the gravities 1.019 as a minimum to 1.090 and 1.092, together with each intermediate rate, and enables any one of ordinary ability and possessing a hydrometric gauge to form a comparatively true idea of the composition of such juices within these limits. From experiments and the information here afforded, it has been determined that the specific gravity should exceed 1,066, representing a percentage of crystallizable sugar present equivalent to 70.48 of the total solids contained in the juice, in order to obtain a satisfactory result, as shown in the remaining columns of the table and explained in the report.

With reference to the analysis of the soils as bearing upon the saccharine production, nothing can be said from the amount of data afforded, as it is too scanty to base any reliable conclusions upon it at this stage, but it must be kept in mind that if sorghum cane is to be cultivated to any extent for the production of sugar it will be due as much to the increased knowledge obtained of the terrestrial and climatic necessities of the plant as to the perfected applications of improved and economic methods of manufacture.

Table 96, showing the "comparative value during the working period of sorghums and corn stalks," is a tabulated expression of the results deduced from the "specific gravity table," and affords the premises upon which Dr. Collier has based his conclusions as presented on the previous page of his report. Without discussing the merits of each variety as therein exhibited, we can safely agree as to the results obtained by him, and more particularly with reference to the eight varieties especially referred to. The "available results," however, must be accepted as a yield produced under fair circumstances and accurate scientific research, and must be valued accordingly as the possible product of a manufacture based on chemical analyses.

Having thus examined the results and methods employed by Dr. Collier, it remains only for us to compare the chemical with the commercial value of the substance, or, in other words, to investigate the probabilities surrounding the conversion of the saccharine equivalent existing in the juice into the commercial product.

This portion of the subject, which is probably the most interesting to the practical sugar-maker, is one which can be treated only in one way—that of analogy. No reliable data beyond results obtained by crude, not to say destructive processes with respect to the production of sorghum sugars, are to my knowledge in existence. Beyond the fact that a crystallizable sugar is attainable, and has been produced by accident rather than intent (a result for many years considered impossible), in greater or less quantities, and in the way of experiment, but little is practically known. That such has been the case is not to the practical sugar-maker a matter of astonishment; on the contrary, the reverse would have been so, in view of the methods followed and the character of the apparatus employed.

The constitution of sugar in a comparatively pure saccharine solution, whether the latter be derived by expression from the canes themselves or formed by the dissolving of the crude material in hot water, as in the refineries, is such that it is readily affected by the surrounding conditions of temperature, atmospheric purity, and cleanliness as regards the containing vessels. In proportion as the attenuation of the solution is increased the action of these influences becomes greater and more rapid, and consequently unfavorable results are frequently obtained, where under proper management a satisfactory return could be easily secured. This is proven by a glance at the average yield of the sugar-canes of the West Indies, where the average analysis of the best canes under favorable circumstances exhibits an average of 18 per cent.; crystallizable sugar in the juice, out of which amount the average best result in the majority of cases of crude sugars does not exceed 6½ per cent. This result is based upon an examination of juices weighing 9°

Baumé, and an analysis of Otaheite cane juice by McCulloh, of the specific gravity of 1.0343, which upon polariscopic examination gave the following result:

```
Water ----------------------------------------------------------------------------  81. 36
Crys. sugar ----------------------------------------------------------------------  18. 07
Ext. matter -----------------------------------------------------------------------   0. 57
                                                                                    ─────
                                                                                    100. 00
```

or 18.07 per cent. of cane sugar. It must, however, be remembered that the planters frequently commence working juice averaging slightly over 5½° Baumé, so that practically speaking the comparison would not be a perfectly just one if regarded as absolutely correct.

With regard to the sugar beet we find the percentage of sugar in its juice to be about 11.30 on the average as generally given, and according to Professor Deherain under certain circumstances to attain a saccharometric strength equivalent to 20 per cent. in special varieties. For the purpose of judging of the adaptability of sorghum juices to the manufacture of sugar I present the following table giving the relative composition of cane, beet, and sorghum juices based on average data, the latter being an average of the twelve first varieties as presented in table 88 of Dr. Collier's report, 1880, p. 110:

	Cane juice.	Beet juice.	Sorghum juice.	
Water	81.00	82.60	82.88	
Cane sugar	18.20	11.30	12.57	
Glucose...	Trace.	....	1.40 }	4.55
Solid matter..	0.80	5.30	3.15 }	

From the above table we may fairly conclude that the comparison develops a greater similarity between the juices of the beet and the sorghum than between the sugar-cane and the latter. Without referring to the comparative equalities of the moistures and cane sugars present in the two latter (beet and sorghum), we may remark at once the greater quantitative approximation of the amounts of solid matter existing in these two substances as being more important from the fact that in proportion to the relative amounts of these solid matters do we find the results as developed in cane sugars reduced. In other words, it is to the amount of solid matters present that we must look for a true measure of the available amounts of sugars that we can obtain, and that in proportion as we are enabled to eliminate or destroy the effect of these elements in the same ratio will we have the power to obtain an increase of saccharine product. Beyond this it is further to be noted that the existence of uncrystallizable sugar in the beet has been claimed by Mr. Bracormot—a fact which would further assimilate its juice to that of sorghum—but denied by Selouze and Pelegot. Its presence has also been claimed to exist in cane juice, according to the analyses of this plant by Icery and Caseca.

In the preceding table while the sugar (crystallizable) existing in the sorghum does not approximate to the amount present in cane juice, it apparently surpasses that of the beet, but the consideration of the presence of glucose in the juice must be regarded as modifying materially its final value. Whether the existence of this substance in the juices of sugar-producing plants can be reduced or eliminated entirely is a matter of interest and worthy of careful examination. That it is capable of reduction seems to be proved by the analysis presented in the case of "Link's Hybrid," in which variety we find the composition to be as follows (see table 12, page 59, Collier's report, 1880), and also from the examination of mature sugar-cane:

Analyses of Link's Hybrid (Collier).

Date.	Specific gravity.	Juice expressed.	Glucose in juice.	Sucrose in juice.	Solids not sugar in juice.
		Per cent.	Per cent.	Per cent.	Per cent.
July 24	1.030	56.90	2.85	3.17	2.18
August 19..	1.066	63.69	1.51	11.86	2.95
September 3..	1.076	64.15	0.88	14.75	3.14
October 22..	1.086	63.36	0.52	16.68	4.99
November 4...	1.084	62.33	0.46	16.43	2.71

In this particular case we find the amount of sucrose present to be nearer that of the average West Indian and Louisiana cane, while a continual reduction in the proportion of glucose also occurs. With juices equal in strength to these, or approximating to such values, and in view of the improvements in the saccharine properties of beet juice, as demonstrated in the increase of their sucroses from an average of 11.30 per cent. to 20 per cent., as shown by Professor Deherain, in the case of samples No. 1 and No. 2, improved beets, No. 348, we may, under the present circumstances, reasoning from analogy, conclude that while we have a sugar constituent existing in the best sorghum juices which will, with careful manipulation, produce profitably a commercial sugar, we have every reason to believe from the data herein afforded that this amount can be very materially increased, and to an extent sufficient to render it under ordinary circumstances a profitable industry in many sections of the country.

In order, however, to attain this end, and to obtain such a maximum result, it mustbe remembered that a careful manipulation of the raw juice is a *sine qua non* to success. A correct knowledge of the composition of the substances which are to be operated upon becomes an absolute necessity, and an avoidance of such processes and methods are as injurious to the operation, and its final result, are equally demanded as requisite elements of success. Perhaps in no stage in the process of sugar-making, whether in the treatment of the true canes or of the sorghums, is so much importance rightly adjudged, and in practice so little care taken, or rather absolute carelessness exhibited, as in the very necessary process of defecation. Practically speaking, in the majority of cases, as far as an actual beneficial defecation is effected, such a result as a perfect one does not exist. The operation as generally performed is done in a very crude manner, and the addition of the defecating material (generally lime), if it be suitable at all for the purpose intended, is usually applied in such a manner as rather to be an actual detriment than an aid to the operation. Frequently a greater quantity is added than the necessities of the case demand, and the apparent failure of it to act as rapidly as expected generally is followed by an increase of the dose, thus converting an actual remedy into a poison. The want of precision in grading the proportion of the neutralizing agent to the chemical necessities of the juice is without doubt one of the greatest faults common among sugar-makers, and yet this is really one of the most important operations of the manufacturer because all the succeeding processes are in a great measure dependent upon it. When once cane juice has been expressed from the stalks, the operations of the sugar-maker should progress without delay, because exposure to the air will, in a majority of cases, produce a viscous fermentation, resulting partly from this cause and partly from the nitrogenous substances held in suspension in the juice. If free acids exist in the juice, lime, in saturating them, produces uncrystallizable salts which tend to keep a portion of the sugar in a soluble condition, and to this cause we can attribute a large proportion of the molasses which is usually formed.

If alcoholic fermentation has set in, the action of the defecating material operates in a different manner. It unites with the carbonic acid, and combines with the glucose which accompanies or precedes its formation; this second product meets with almost immediate decomposition, the glucose going to form glucic acid and, with the lime, glucate of lime. This operation does not stop at this point, but still progresses, and this salt now becomes changed into a new substance—probably the molassate of lime—which forms the coloring matter present in brown sugars and molasses. This latter substance is, however thought to be less injurious in its effects than glucose itself, for the reason that it is less viscid than the latter, and although it destroys a certain amount of saccharine matter, does not prevent crystallization to the same extent as glucose.

Defecation, therefore, should be a neutralization of the acids as nearly as possible, but in practice we find the actual result to be slightly alkaline. If alteration has taken place owing to various causes, as previously shown, glucate of lime is the result and saccharate of lime if otherwise. This latter is an uncrystallizable salt, which, when present in quantities, owing to its greater decomposing action, also renders evaporation slow and difficult, and opposes to some extent crystallization, according to the best authorities on the subject.

The difficulties here referred to may, perhaps, be ameliorated to some extent by passing the fresh juices at once into a tank or receiver to which a low degree of heat can be applied, and in which the action of specific gravity may aid in separating the heavier feculencies present, then passing the partly clear juice through a thin layer of bone-black in order to separate mechanically as much as possible the remaining feculencies, and especially the gummy and viscid substances in the juice. Thence it goes at once to the defecators. where it is to be treated with lime, preferably graystone lime, freshly slaked and diluted with water to the consistency of milk. The practice of using dry lime, or of mixing it with a portion of the juice, is not one to be recommended, as it produces, with the mixing juice to the amount that is employed, the very result that should be most guarded against. After a perfect defecation it next passes to the clarifiers, or may be passed through bone-black which has already been used for bleaching, which is the preferable method, as the action of the black extracts the excess of lime present in the juice, which has the double tendency of producing coloration in the juice on the application of heat and the formation of *cal* or scale in the vacuum-pans. Powdered bone-black may also be used with satisfactory results in the course of defecation, as also albuminous substances. Excess of lime may also be neutralized in many ways; and while alum and cream of tartar have been suggested for this purpose, sulphuric acid is to be regarded as being preferable to either when the after results are taken into consideration. The indications of a perfect defecation are easily distinguishable by the practical sugar-maker, and too well known to need description here. Clarification, which is simply a supplementary defecation without the intervention of anything except heat, needs no comment as being based upon the same general principles that control the previous process of defecation. We may therefore pass at once to the consideration of the next stage in the process, that of evaporation, and the means used to effect this object.

The apparatus employed in the evaporation of saccharine solutions may be divided generally into three classes: firstly, those in which direct (fire) heat is applied, and of which the Jamaica or open kettle train is the representative, and probably the best of its class, together with the various modifications of "patent" fire evaporating pans more aptly styled by Dr. Ure "sugar frying-pans;" secondly, open steam trains or apparatus in which the evaporation is produced by the radiation of steam heat or hot air in the presence of the atmosphere or under ordinary pressure; and thirdly, vacuum apparatus, or such in which evaporation is carried on at a point below the normal pressure of the atmosphere and at a reduced temperature.

As it has been with apparatus of the first class that most of the experiments made in the concentration of sorghum juices have been performed, a few remarks on this special class may not be out of place at this stage of the present inquiry. According to Hochstetter, during the evaporation of cane juice in open trains, wherein the liquid is exposed directly to the action of the atmosphere at a high temperature, formic acid is liable to be produced, and this formation is claimed by him to exert a most pernicious effect upon both the color and crystallizable powers of the cane sugar. It is also remarked by the same authority, that a similar effect occurs during the process of claying, by the direct contact of the large surface exposed by the heated drops with the air, and the action produced thereby. The production of caramel and of glucose is also a direct result of this method, and for weak solutions and those of the character of the sorghums cannot for these reasons be for a moment considered.

The next class—those involving the use of open trains using either steam or hot air as evaporating agents—while they are an improvement upon the former, fall far behind in point of effectiveness the results to be derived; for although a better control of the temperature is afforded than by the use of direct heat, the difficulty of a necessarily high temperature at the termination of the process destroys to a very great extent whatever advantages may have accrued during the primary stages of the evaporation. That a portion of these ill effects may be materially obviated in this class of evaporators is evidenced by the practice followed by the beet-sugar manufacturers, who in some cases use wooden covers and steam chimneys for their evaporating vessels. The amelioration referred to is caused by the exposure of the surface juice to a covering of steam, whose effects are less injurious than those of heated air. This modification of the open evaporator is to be recommended, as being one involving but little expense, and would no doubt if suitably arranged give comparatively satisfactory results, independently of effecting a greater economy in the matter of fuel.

The third or vacuum apparatus as invented by Howard, and as modified by various manufacturers, must however be regarded as the most perfect system yet applied to the manufacture of cane or other sugars. Filling as it does all the conditions requisite to a favorable attainment of the end in view, it is without doubt the only proper method worthy to be employed in exact investigations of this character, and when combined with steam defecators, bone-black filters, and mechanical filters, and erected as a double or triple effect apparatus, we have presented to us the most perfect combination of mechanical devices that has yet been invented for the production of sugars.

While comparatively cheap as regards cost, both in the first instance and with respect to their operation, competition among manufacturers has in many instances produced an inferior machine; and a short-sighted desire to economize on first cost has led to the construction of apparatus which, while a great improvement over the open trains, are in some respects inferior to what they should be. In this category we may place vacuum-pans without steam-jackets, which rely for their heating surface entirely upon the amount of worm contained in them, as being probably the worst of their class and a source of endless annoyance to the sugar boiler, not to say loss to the manufacturer. A substitute for the steam-jacket has been sought for by placing the lower worm close to the bottom of the pan, but this device does not seem to meet with satisfactory results in practice. Where economy of steam is a matter of great importance the use of a single pan, arranged in direct connection with a steam recipient so as to utilize the exhaust steam of the pumps and engines, may be successfully employed, and the evaporation carried on in the open steam or kettle train up to a point varying between 18° and 26° Beaumé, but in this case large capacity is required both in the case of the vacuum pump and pan.

The process of crystallization is the next that claims our attention, and may be regarded rather as an effect produced by the continuation of the previous operation, than as being separate and distinct from it. The methods pursued to obtain this have for their basis the well-known fact "that sugar is soluble in water in a ratio proportional to its temperature up to the degree of 270° Fahr., at which temperature the result is a sirup so exceedingly viscid that if it be allowed to cool it will be found that the sugar cannot crystallize in the usual manner, but simply solidifies in the form of a transparent amorphous mass, similar to that prepared by confectioners and popularly known as candy. While, however, crystallization ceases at this point, it is to be remarked that there still remains a certain amount of moisture, sometimes called the water of solution, united in the proportion of 10 per cent. of water to 90 per cent. of sugar. If now the evaporation be pushed to a point beyond this, so as to perfectly desiccate the mass, the increased temperature required to do this operates in such manner as to effect a complete decomposition of the sugar itself, and thus sets the bounds of the possible limits of this process, as far as heat is concerned. On the other hand we find that while sugar is soluble in one-third its weight of cold water at the temperature of 60° Fahr. a saturated solution contains 66 per cent. of sugar and 33 per cent. water, while at 212° Fahr. by reason of the increase of solubility due to the change of temperature the same condition indicates the relations of 83.33 per cent. of the former to 16.66 per cent. of the latter, or the proportions of 5 to 1. For this reason, if we undertake to evaporate a solution which is saturated at the temperature of 60° Fahr. and to so apply heat so as to maintain a constant temperature of 212° Fahr. it will be found necessary to drive off in the form of vapor one-half of the water contained in the solution before the latter will have become fully saturated at the higher temperature. In other words the proportion of water must be reduced from one-third the total amount present at 60° to one-sixth at 212'. As soon, however, as the saturation point is reached any further progress in this direction will cause the sugar to separate in a crystalline form on account of the solvent power of the water present having been exceeded, and this process will continue until the deposit becomes too thick either to transmit the heat readily or allow of its convenient passage through the exit valve of the pan.

Evaporation under atmospheric pressure, therefore, exhibits results in accordance with the indications shown in the previous paragraph and applies to the various forms of open apparatus used in the art, and in such others wherein an equivalent surface pressure is maintained during the operation. We encounter, however, very different phenomena, if we eliminate the pressure of the atmosphere and perform the operation in a vacuum. Subject to this condition, we find that the operation is performed under a very reduced temperature, and in consequence of the rapid absorption of latent heat the "mother liquor" or unevaporated portion of the sirup will necessarily contain a lesser amount of sugar. Taking 175°-180° Fahr. as the limit of temperature we can estimate, for all practical purposes, with sufficient accuracy that water will dissolve four times its own weight at this heat, being then fully saturated. Further evaporation, therefore, necessarily causes the deposition of crystalline sugar as in the former case. In practice the operation is carried at various temperatures, ranging from 120° Fahr. up to 180° Fahr., determinable by the character of the sirup acted upon and the perfection attained in the previous processes. In boiling for grain, as it is technically termed, the same principle maintains, the skill of the sugar-maker being directed to form primarily a series of nuclei and to afterward erect a superstructure upon them without permitting the formation of a secondary series. It is in this way the higher grades of sugar are now usually obtained, and to this method and the application of centrifugal force in the drying apparatus are due the existence of the well-known class of centrifugal sugars. Muscovado sugars, however, follow a different method, the sirups being concentrated to a lesser degree of saturation, the final crystallizing action taking place in open tanks or coolers under atmospheric conditions and at a temperature contained within the limits of 95° and 105° Fahr. Here, too, the separation of the solid sugar follows the same general law of saturation, and in proportion to the length of time employed in the tanks in cooling will the size of the crystals formed therefrom depend. Molasses sugars of strong test are boiled so as to form minute crystals in the pan, but are dropped into tanks and the operation is concluded in the same way as in the case of Muscovados.

The treatment of soured or frozen canes is a matter involving considerable interest to cane-growers both in Louisiana and other parts of this country, and owing to the fact of the adaptability of the sorghum to higher latitudes must necessarily be important to the producers of this plant.　It may be briefly stated that while much is open to the investigation of the chemist in this department, practical experience has indicated the profitable crystallizability of such juices, when improved apparatus and proper conditions have been employed.　It is also to be observed that while such canes may be worked in the ordinary way, a larger percentage of crystalline sugar is produced when methods similar to those pursued in the course of extracting sugar from West Indian and other molasses of strong polariscopic test by the northern refineries are followed.

Of the centrifugal machine and its numerous adaptations to the draining of various classes of sugars, little is necessary to be said, as its common use among sugar-makers and refiners has rendered it either in one style or another generally known.　At this point, before entering upon a consideration of the conclusions to be drawn from the previous investigations, it may be well to compare the results presented to us by McCulloh of the workings of the various apparatus as compared with each other in the production of sugars from Louisiana canes.

In the following table the comparison is made between the several different systems of manufacture as therein given, both with respect to the amounts of sugars produced, the relative quality or grade of product, and the total pecuniary value in a relative point of view of each result, the whole being based upon an equal amount of extractable sugar in each case.

Value of a crop of cane made into sugar by six different processes, into hogsheads, without being siruped, at the price each class has been sold or is worth this season, 1847.

[The whole quantity of dry sugar being in each case 653,367 pounds and the boiling power required for each crop being 8,000 pounds.]

Method.	Pounds of first sugar.	Value per pound in cents.	Total value of first sugars in dollars.	Pounds of second sugar.	Value per pound in cents.	Total value of second sugars in dollars.	Gallons of molasses.	Value per gallon in cents.	Total value of molasses in dollars.	Total value of crop.
1........	433,000	4½	19,485	*43,300	2	866 00	26,734	18	4,812 12	25,163 12
2........	433,000	4½	19,485	153,500	2⅔	1,221 25	13,857	18	2,494 26	26,200 51
3........	433,000	5	21,650	162,000	3½	5,265 00	12,874	18	2,317 32	29,232 32
4........	433,000	4	17,320	153,500	2½	3,837 50	12,857	18	2,314 26	23,471 76
5........	433,000	5½	23,815	162,000	3½	5,670 00	12,871	20	2,574 20	32,059 20
8........	440,000	6½	28,600	163,000	4¾	7,742 50	11,949	20	2,389 80	38,732 30
+8......	478,500	6	28,710	141,000	4½	6,345 00	10,041	20	2,008 20	37,063 20

Comparative expenses, consumption of fuel, profits, &c.

	Plantation expense.	Bone-black expense.	Wood burned, cords.	Saving or loss, in dollars.	Net proceeds.	Relative gain, by dollars.	Cost of apparatus.
1..................	$7,000		1,515		$18,163 12	$13,440 43	$2,000
2..................	7,000		1,515		19,200 37	12,403 18	2,500
3..................	7,000		1,515		22,232 32	9,371 23	4,000
4..................	8,000	200	1,948	L. $974 25	14,477 37	17,126 18	12,000
5..................	8,000	200	1,948	L. 974 25	22,885 55	8,718 00	12,000
8..................	8,000	200	550	S. 2,171 75	31,603 55	....	10,000
+8................	8,000	200	550	S. 2,171 25	31,034 45	569 10	10,000

(In the foregoing table taken from M'Culloh's report, method 1 represents "the old set of kettles;" 2. Set of kettles for sirup, strike high pressure steam pan; 3. Set of kettles for sirup, and strike vacuum-pan; 4. Open high pressure steam pans for sirup and strike; 5. The same for sirup with strike vacuum pan; 8. Rellieux's triple and quadruple effect pan apparatus, clarifiers, and filters; +8. The same with results obtained from high boiling).

In the above table we have the relative values of each of these methods, as far as production, value of the result, and expense attending the same are concerned, together with the relative cost of each apparatus at the date of the compilation of this table.

These results are alike both with reference to the total amount of gross values produced, and the relative economy of each method with respect to cost of production referred to the net product. Thus in both instances they stand as follows: 8, +8, 5, 3, 2, 1, 4; the method 8 giving the best and 4 the poorest, proving that the rules indicated by scientific investigation are substantiated by the results of practical experience. From these data we may conclude, therefore, that vacuum apparatus when properly combined with suitable clarifying appurtenances will always give the most satisfactory returns and in the increased value of the product will in a short period repay the original first cost.

Having thus compared fully the chemical constituents of the canes under discussion, the processes best adapted for the attainment of the ends in view, together with such suggestions as may seem proper towards enabling the operation to be conducted with a minimum loss, we may return to the comparison of the actual results obtained in practice, with the "available" ones presented by Dr. Collier in his tables. If we accept it as a fact that Louisiana cane will produce on an average 2,000 pounds of sugar and 120 gallons of molasses to the acre (and we believe that taking the plant and ratoons together this will be found a high estimate) we have the following data:

Average Lousiana cane:
    Sugar _____ pounds__ 2,000
    Molasses _____ gallons__ 120
French beets:
    Sugar _____ pounds__ 3,600
    Molasses _____ gallons__ 156
Sorghum cane:

    Sugar (average) _____ pounds__ $\left\{\begin{array}{l} 1,417 \\ 2,374 \end{array}\right.$

    Molasses (estimated) _____ gallons__ $\left\{\begin{array}{l} 90 \\ 80 \end{array}\right.$

AGRICULTURAL EXPENSE.

Average cost of working per acre:
    Louisiana cane (estimated) _____ $14 00
    French beets_____ 14 00
    Sorghum cane_____ 11 50 to 17 50

VALUE OF RESULTS.

Louisiana cane:
    Sugar_____ $140 00
    Molasses _____ 72 00
                                                                   _____
                                                                   212 00
                                                                   ========

French beets:
    Sugar_____ 189 00
    Molasses _____ 16 38
                                                                   _____
                                                                   205 38
                                                                   ========

Sorghum cane:

    Sugar (average)_____ $\left\{\begin{array}{l} \$113\ 36 \\ 54\ 00 \end{array}\right.$
                                                                   _____
                                                                   167 36

    Molasses (estimated) _____ $\left\{\begin{array}{l} \$189\ 72 \\ 48\ 00 \end{array}\right.$
                                                                   _____
                                                                   *237 72

From the above data, which is approximated as closely as the information attainable in this regard will allow to actual results, we can conclude as follows, independently of the final determinations reached in Dr. Collier's report:

1st. That sorghum canes as a source of sugar production possess on an average less crystallizable sugar in their juices than those of the average sugar-cane, but more than that contained in the juices of the average French beet.

2d. That a comparison of the analyses of the juices of the beet, the sorghums, and the

---

* Mr. Johnson's estimate omitted the value of the seed of sorghum, an item which it is important to state, and which in the opinion of many cultivators is fully equal to the cost of cultivation or more. Evidence on this point abounds in this report.—*Com.*

true cane, seems to develop the fact that a greater similarity, chemically speaking, exists between the sorghums and the beet than between the latter and the sugar-canes, the quantity of "solids not sugar" (ash) in each case being regarded as a factor of greater importance in the matter of sugar production than the other features.

3d. That the capacity of the sorghums for improvement, and their consequent increase to produce crystallizable sugar, is evidenced by the superior results obtained from particular races and gives us strong reasons to believe that it may be possible to approximate its yield in the near future to that of the ordinary sugar-cane.

4th. That owing to the character of the plant and its similarity in this respect to maize, its relative cost of cultivation is necessarily less than that of the beet, and taking into consideration the element of time with reference to the period of its growth, less than that of the sugar-cane. It is also better adapted to the general conditions of soil, climate, and system of agriculture pursued in the United States than that of the beet.

5th. That a fair trial and effort to determine the practicability of this industry can only be made by the use of the best class of machinery, under the management of thoroughly competent men of practical ability, before any decision in regard to the results can be rendered upon this matter, because the profits to be derived from the cultivation and manufacture of sugars depend rather upon the relations existing between the actual amounts of sugar extracted and converted into the commercial product and the total sugars in the plant, than upon the latter alone. Thus, while in the case of the ordinary cane a result of $6\frac{1}{2}$ per cent. out of a total of 18 per cent. is the common result, a product of 7 or 8 per cent. out of a total of 14 per cent. would be a better one in this particular instance, independent of the relative increase, inasmuch as the article produced would from its quality be of more value, over and above the fact of its involving a less cost in the matter of cultivation.

6th. From the conclusions drawn in the body of this communication, as also from those presented in No. 2, it follows that the methods best adapted to the treatment of sorghum juices should lean rather towards those employed for the beet than the ones in use for cane, for the reason that the difficulties in these juices are similar and arise in great measure from the same sources.

7th. As an element involved in the practical success of this industry, the relative value of sorghum cane as compared with those of other crops of the sections in which it may be raised, viz, wheat, rye, oats, corn, &c., becomes important, for the reason that the cost of the raw material and its adjuncts influences to a very great extent the value of the final product, because no agricultural industry can ever succeed unless the monetary value of the product exceeds that of the one it supplants.

In conclusion, I would state that while exceeding the limits originally intended for this communication, and perhaps digressing on details not directly connected with the sorghums themselves, the fact of the very intimate relations between the three classes referred to must be my excuse. As a practical sugar-maker, however, my own experience of many claiming to be such, and their constantly repeated errors has led me to be more diffuse on the subject of manufacture than I should otherwise have been, for the reason that if these conditions exist among such men we must necessarily expect to find them in a greater degree present among those not possessing any experience—such as the ordinary farmer.

Let us hope that what chemistry in the matter of the sorghums promises us in the present may be brought to a successful realization in the near future, and that following in the footsteps of its older brother it may meet with final success.

Very respectfully submitted,

C. CONRAD JOHNSON.

B. SILLIMAN, *Chairman Sorghum Sugar Committee,*
National Academy of Sciences, Washington, D. C.

28.—*LETTER FROM HENRY B. RICHARDS, OF TEXAS, RESPECTING ORANGE CANE, PERENNIAL, BY HIS EXPERIENCE.*

[Referred to in the report, p. 28.]

LA GRANGE, TEX., *September* 25, 1882.

DEAR SIR: I have been working with improved varieties of sorghum, on a small scale, since 1875 until the summer of 1880. I considered until then the Early Amber the best for all purposes. In the drought of that summer it fell far behind the Early Orange (sent me that spring by Mr. Hedges, of Saint Louis,) and it (the Amber) failed entirely in making a second crop that season. The Orange, on the contrary, made a good average second crop.

Again, in 1881, with the severest drought I ever saw, Amber was a failure; Early Orange made two crops, nearly average. I now plant the Orange only. Two mature crops of it, for the mill, are certain, and, for forage, three crops.

In regard to sugar, my works are too small for profitable results in that line, although I have made, experimentally, 200 and 300 pounds of sugar during a season.

Owing to the scarcity of and demand for pure sirup in this section, I have not kept a single barrel of sirup this season long enough for it to granulate much, and have only boiled to 228° Fahr., sirup standard. Specimen samples of sirup that I kept have all granulated more or less, notably one, made June 19, which is three-fourths solid sugar, in very large grains. My sirup is all contracted for by one merchant at 55 cents per gallon at mill, for all I can make during 1882. I have made near 2,000 gallons to date, and expect to work for two months yet. The yield this season per acre is not satisfactory. Juice is pretty rich, ranging from 9° B. to 12° B., mostly the last; but the stand is bad generally, and cane low and branching at the top to three and four, sometimes more, heads, which being cut off, leaves the cane very short, say 4 to 6 feet, for the mill.

The average yield for my mill per acre is about 65 gallons; some have made 80 and 90 gallons; some as low as 40.

With regard to the Perennial Orange, its yield was 86 gallons per acre; juice 11° B., first crop, and a very poor stand, as it was totally unprotected during the winter. The second crop from it for this season will soon be ripe.

I intend protecting my stubbles this winter with two furrows thrown on them, and confidently expect a much earlier and heavier crop next year.

I find the juice from the old stand richer by 2°, under the same circumstances, than seed cane, and the growth of stalk more vigorous and rapid; it also matures earlier.

No other variety, except the Early Orange, possesses this perennial quality. The plant has no insect enemies that I am aware of; it flourishes when other crops are parched with drought. By far the most valuable forage plant in existence, yielding three crops per annum, each crop giving three times the amount of fat-producing food that the same area of corn will give. The only drawback to its culture I know is the difficulty of procuring labor during the season of cotton picking.

Very respectfully, yours,

HENRY B RICHARDS.

Professor SILLIMAN, *Chairman Sorghum Sugar Committee, N. A. S.*

P. S.—I consider the two *certain* crops of this Orange cane a much safer and better investment every way than any one crop of Ribbon cane, which is the most *uncertain, unreliable* thing on earth.

29.—LETTER TO THE CHAIRMAN FROM GENERAL ASHBEL SMITH, M. A., M. D., ETC., HOUSTON, TEX.

[Dr. Smith's remarks on the peculiarities of sorghum grown under the latitude of Texas are of special interest. He is one of the oldest and most intelligent agriculturists in Texas, where he has passed a long life, varied by his diplomatic and other public duties, having been minister to France and Great Britain from the Republic of Texas before its annexation to the United States. His estimate of the biennial or perennial character of sorghum in that latitude is interesting, and more so are his remarks on the successive crops, in the same season, from one early sowing, and its vigor against frost destructive to maize. By Dr. Smith's statement sorghum is a good sugar crop in Texas for six months, from June to December, and hence the long maturing varieties, like Honduras, Mammoth, &c., are of special interest.]

EVERGREEN, NEAR HOUSTON, TEX.,
October 18, 1882.

MY DEAR SIR: Your esteemed favor of September 19 reached me a few days since, after an absence of some time from home.

I am afraid I shall not be able to give any information valuable or new to you concerning sorghum. I can only evince my disposition to do so.

I have cultivated a patch of sorghum, an acre or such a matter, every year for several years, not as a leading crop, but as feed for hogs and green forage for horses and mules.

I have long known that sorghum is a biennial, perhaps a perennial. This, however, is an inconsiderable advantage, or rather none at all. For annual planting is of less labor than cleaning the ground around the ratoons; besides, annual planting leaves the ground in better condition. The cost of seed for planting is practically nothing, for a nickel's worth is enough to plant an acre. The sorghum of the sugar species known to me is a very hardy plant, and bears without injury a degree of cold which completely kills Indian corn. Consequently it may be planted very early, with us in Texas *early in February.* It is in another respect hardy, for when fairly started to grow it grows so rapidly as soon to outstrip nearly all the weeds of a field, making it a crop of easy culture. It suffers as little from being bruised in working as Indian corn. So far as I know it has not in Texas suffered from any disease, nor is it yet infested by any insect. In Texas and, I presume, in more northern climates, it tillers very numerously. I have already mentioned its very rapid growth. When cut down to the ground it sends forth shoots so promptly, and these grow so vigorously that two crops of matured canes, ready to feed or grind for sirup, can be cut from the same plants and their ratoons, and two or three crops of well-matured seed gathered from them during the same season.

In Texas we have several varieties of sorghum, between which the chief difference is of those which bear black seed and those which bear red seed. My own observation does not enable me to speak of their relative worth. You are, of course, aware how readily the sorghums hybridize with each other and with their congener, broom-corn.

In Texas sorghum is often rolled for sirup on a small scale, but not for sugar, though we know that sugar can be extracted from it. A sufficient reason for the latter fact is that the West India cane, *Saccharum officinarum,* flourishes well here. A neighbor of mine makes every year from sorghum a few barrels of sirup, a fair article, better than the manufactured sirup of glucose, but decidedly inferior to the sirup from the proper sugar-cane. The sorghum sirup, which I have tasted, leaves in the mouth a pungent sensation—an avoidable or removable defect, for it does not exist in the juice of the mature sorghum cane.

An acre of sorghum from one cutting yields much less saccharine matter than an acre of West India sugar-cane. This disparity in yield may be compensated or reversed in favor of sorghum by the successive crops or cuttings of sorghum from the same roots in one season before adverted to.

It is certainly a great advantage of sorghum that *rolling the cane may be commenced in June and continued without intermission until Christmas.*

The rapidity with which saccharine juices take on fermentation, especially in warm weather, would in such case necessitate the use of sulphurous antiseptics to secure the immediate commencement from the mill of their conversion into sirup or sugar.

Sugar-making is a good deal more than simple boiling. It requires skill and practical tact. These are readily acquired under an experienced sugar-boiler. To ascertain the reliable capability of sorghum as a source of sugar supply, get an experienced, practical sugar-boiler from Louisiana or Texas. Then place the product in the hands of one of your northern sugar refiners.

I surely can see no reason why sorghum cannot be cultivated successfully in districts so far north of the *Saccharum officinarum* belt as *to furnish the entire supply for consumption.* It must have some advantages over the sugar-beet of Europe. It has less importance for us in the great Mexican Gulf districts of Texas, for the *Saccharum officinarum* flourishes and matures sufficiently throughout all these districts.

On reading over what I have written I am surprised a little at the meager details of this letter. It would be indeed a sincere pleasure to add any facts, if I knew any, in addition to those you already possess.

Faithfully, your friend,

ASHBEL SMITH.

Professor SILLIMAN, *&c.*

30.—*DUPLICATE ANALYSES OF SORGHUM JUICES.*

[See report, p. 26.]

For the purpose of controlling the results of analyses, there have been made during the season of 1882 twenty-four analyses of sorghum juices in duplicate at the Department of Agriculture.

In no case did those who were engaged in the analyses have any reason to suspect that they were at work upon duplicates, the samples having been prepared and sent in to the laboratory under their several numbers, as being individual specimens of juice. Thus Nos. 105 and 113 were duplicate juice, and so are Nos. 107 and 115.

It will be observed that the agreement is quite as close as could be expected in work of such a character, and that the average results given at the close of the table show that in the analytical work there is nothing to cause doubt as to the substantial accuracy of the work recorded.

Duplicate analyses of sorghum juices.

Number of analysis.	Specific grav.	Glucose.	Sucrose.	Solids.	Polar.
105	1042	5.40	2.72	1.69	3.06
107	1034	3.25	3.73	1.30	4.24
109	1038	3.27	4.69	1.29	5.28
111	1053	3.38	8.30	1.14	8.01
133	1072	2.10	12.96	3.07	13.44
158	1068	1.15	11.79	3.26	12.06
160	1063	2.79	10.06	2.55	10.21
161	1072	.93	13.31	3.28	13.60
168	1057	3.23	8.37	2.35	8.68
169	1056	2.02	8.92	2.63	9.36
173	1068	1.57	12.03	2.64	12.44
182	1057	1.48	8.48	3.41	8.54
189	1058	2.13	10.41	2.50	....
191	1057	3.29	8.50	2.24	....
192	1069	2.44	11.65	2.79	....
194	1056	2.42	8.86	2.24	9.01
206	1038	.67	5.72	2.92	....
224	1071	1.05	13.15	3.27	13.44
214	1069	1.89	12.48	2.73	12.91
216	1073	1.82	12.94	3.20	13.68
212	1071	.84	13.39	2.98	13.53
220	1071	1.19	12.54	3.40	13.23
221	1061	4.11	8.22	2.81	8.70
219	1061	1.64	10.31	3.24	10.93
	1435	54.06	233.63	60.43	195.34
Average	1059.8	2.252	9.729	2.627	10.281

Number of analysis.	Specific grav.	Glucose.	Sucrose.	Solids.	Polar.
113	1042	5.65	3.13	.87	3.20
115	1034	8.26	3.83	1.25	3.72
114	1038	3.27	4.86	1.21	7.24
116	1053	3.41	7.82	1.57	7.90
136	1073	1.93	12.81	3.27	13.53
160	1069	1.16	11.59	3.03	12.14
157	1063	2.92	10.23	2.21	10.28
163	1073	.93	13.31	3.54	13.60
180	1057	3.18	8.26	2.29	8.69
181	1056	2.05	8.78	2.45	9.34
179	1068	1.53	11.96	2.70	12.55
183	1057	1.49	8.45	3.09	8.52
202	1059	2.18	9.99	....	10.48
203	1056	3.29	7.91	2.75	8.47
204	1069	2.41	11.60	2.88	12.05
205	1056	2.46	8.70	2.55	....
207	1038	.67	5.93	2.88	....
226	1071	1.12	13.37	3.26	13.48
227	1069	1.87	12.73	2.58	12.02
228	1073	1.54	13.50	2.76	13.76
229	1071	.89	13.73	2.98	13.48
230	1071	.93	11.11	5.15	13.48
231	1061	4.04	8.49	2.54	8.64
232	1062	1.68	10.95	2.56	11.02
	1439	53.86	233.04	60.37	194.59
Average	1060.0	2.244	9.710	2.625	10.242

31.—THE CONNECTICUT AGRICULTURAL EXPERIMENT STATION—ANALYSES OF SORGHUM SEED.

[Annual Report for 1881. Professor S. W. Johnson, Director.]

1. Seed of Minnesota Early Amber cane, from E. M. Dunn, Grafton, Mass.
2. Sorghum seed, from E. D. Pratt, West Cornwall, Conn.

COMPOSITION.

Constituents.	Air-dry.		Water-free.	
	Minnesota Early Amber.	Sorghum seed.	Minnesota Early Amber.	Sorghum seed.
Water..	15.04	16.76	....	....
Ash	1.73	2.17	2.04	2.60
Albuminoids or protein.................	8.13	7.69	9.57	9.23
Crude fiber	1.94	3.21	2.28	3.85
Starch, sugar, and gum, by difference...........	69.65	66.81	81.98	80.30
Fat...	3.51	3.36	4.13	4.02
	100.00	100.00	100.00	100.00

No determinations of the digestibilty of sorghum seed have been reported. Its composition is quite similar to that of the ordinary cereal grains, and it is to be anticipated that it will prove equally digestible and nutritious.

Also the following analyses by Dr. Peter Collier, Chemist to the Department of Agriculture:

Constituents.	Early Amber.	Liberian.	White Mammoth
Water	10.57	9.93	....
Ash	1.81	1.47	1.85
Fat..	4.60	3.95	4.58
Sugars	1.91	2.70	2.58
Albumen insoluble in alcohol..........	2.64	2.64	2.95
Albumen soluble in alcohol.............	7.34	6.90	7.95
Gum	1.10	.72	.95
Starch......................................	68.55	70.17	77.46
Fiber.......................................	1.48	1.52	1.68
Total.......................................	100.00	100.00	100.00

32.—COMMISSIONER LORING'S CIRCULAR.

UNITED STATES DEPARTMENT OF AGRICULTURE,
Washington, D. C., June 6, 1882.

To the manufacturers of sugar from sorghum, beets, and other sugar-producing plants in the United States:

Congress in the appropriation for this Department, for the fiscal year commencing July 1, 1882, has provided for "experiments in the manufacture of sugar from sorghum, beets, and other sugar-producing plants."

In view of the experiments which have already been made at this Department, I have determined to institute the following plan for the coming season, in obedience to the act referred to.

Provision has been made for continuing the chemical analyses of sorghum at the laboratory of the Department, should this be deemed necessary, in order to add to the information already obtained by investigations not only here but also in the agricultural colleges of this country.

On assuming the duties of my office in 1881, I found 135 acres of sorghum containing 52 varieties which had been planted in Washington for the use of the Department. On being informed that the time had arrived for manufacturing sirup and sugar, I engaged the services of an expert in sugar-making who had been highly recommended for the position of superintendent, and operations were commenced on September 26 at the mill, erected by my predecessor, on the grounds. These operations were continued with slight interruptions until the latter part of October, at which time the supply of cane became exhausted. Forty-two acres of the crop were overtaken by frost before being sufficiently ripe for use, and this portion of the crop was so badly damaged as to be unfit for manufacture. The yield of cane, per acre, on the 93 acres gathered was two-and-a-half tons; the number of gallons of sirup obtained was 2,977; and the number of pounds of sugar was 165. The expense of raising the cane was $6,589.45; and the expense of converting the cane into sirup and sugar was $1,667.59—an aggregate of $8,557.04.

The manufacture of sorghum at the Department therefore has been found to be so expensive and unsatisfactory, that the work can evidently be better conducted elsewhere. To repeat the experiment of last year would be unwise under any circumstances, and it is made doubly so by the impossibility of procuring the sorghum cane at any reasonable price in this neighborhood, after the discouraging crops of last year, and by the additional fact that the appropriation is not available until too late in the season for planting to begin.

While therefore such scientific investigation as is deemed necessary at this Department will be continued, the experiment of manufacturing can better be conducted by those who have thus far furnished us all the valuable information we have; and this work I refer to the manufacturers themselves, to whom I submit the following proposition.

Each manufacturer is requested to submit an account of his work to this Department, covering the following points, viz:

1. An accurate account of the number of acres of sorghum brought to his mill; the number of tons of cane manufactured; the yield of sorghum per acre; the mode of fertilizing; the time of planting; the time required for maturing the plant; and the value of the crop as food for cattle after the juice has been expressed.

2. The amount of sugar manufactured; the amount yielded per ton of cane; the quality of the sugar; the amount of sirup manufactured; the process of manufacturing; the machinery used; the success of the evaporator, the vacuum-pan and the centrifugal in the work of manufacturing.

3. The number of hands employed in the mill; the cost of fuel; the cost of machinery; the wages paid for labor; and the price of sorghum at the mill if not raised by the manufacturer.

The returns when received will be submitted to a competent committee for examination, and in order to compensate the manufacturers for the work of making these returns, I propose to pay for the ten best returns the sum of $1,200 each—the decision to be made by the aforesaid committee. Each return must be sworn to before a competent officer.

SUGAR BEETS.

I have distributed to ninety persons a supply of the best sugar-beet seed which I could obtain; and I would request each person having received this seed to send to this Department a statement of the amount of land planted by him; the yield per acre; the fertilizers used; the value of the crop in the market. I also request each person making this experiment to forward to this Department a sample of the crop for analysis. The directions for this will be issued hereafter. An accurate statement of the process of manufacturing beet sugar in this country is of great importance, and I propose to compensate the manufacturers for preparing such statement by the payment of the sum of $1,200 for each of the two best returns submitted to a committee as in the case of sorghum.

OTHER SUGAR-PRODUCING PLANTS.

The promise of 1,000 pounds of cornstalk sugar per acre, which was made in 1841, and has often been repeated with great confidence, both at the expense of the corn crop and in addition to it, not yet having been fulfilled in manufacture, the experiments not having been satisfactory, and the business not having been followed up, it is not deemed necessary to institute sugar-making experiments in this direction during the present year. The same may be said of many esculents which have been classed as sugar producers.

All proposals to enter upon this work for the Department must be laid before the Commissioner on or before August 1, 1882.

GEO. B. LORING,
Commissioner of Agriculture.

33.—*REPORT FROM PROFESSOR SWENSON, DETAILING THE RESULTS OF THE SEASON'S WORK AT THE UNIVERSITY OF WISCONSIN.*

[N. B.—The following report of Prof. Swenson arrived after the Academy's report had been transmitted to the Commissioner of Agriculture. But it is deemed of sufficient importance to form a supplement to the data previously submitted on pp. 34, 35.]

This document is interesting especially from the evidence it offers of the little effect obtained by the use of fertilizers upon the sugar product of the crops treated by different manures. The effect appears to have been rather to diminish the sugar output, as compared with no manure, so far as can be judged from the figures.

UNIVERSITY OF WISCONSIN, AGRICULTURAL DEPARTMENT,
Madison, Wis., November 22, 1882.

DEAR SIR: I forward you to-day a short report of the work done here this season. I am sorry that it comes so late and hope it may reach you in time. I also forward a sample of sugar from plot No. 2.
Use my report as a whole or in part, as you may see fit.
Very respectfully,

M. SWENSON.

Prof. B. SILLIMAN.

DEAR SIR: I take great pleasure in submitting the following brief report of the work done with sorghum canes on the farm of the University of Wisconsin during the past season.
Twenty-six varieties of cane were grown on the university farm during the past season, some of which were from seeds kindly sent by Dr. Collier. The following table shows the results of my examinations:

Variety of cane.	Average weight of stalks.	Stage of the seed.	Per cent. of cane sugar.	Per cent. of glucose.
	Lbs. Oz.			
Chinese No. 1	8	Dough	7.28	1.76
Chinese No. 2	8	Milk	7.71	1.88
Chinese No. 3	14½	Milk	7.26	1.78
Chinese No. 4	9½	Dough	7.80	1.78
Chinese No. 5	9	Dough	7.15	2.08
Chinese No. 6	1 4	Milk	4.18	1.61
India No. 1	1 6	Hard dough	9.45	1.65
India No. 2	14	Hard dough	9.26	1.64
Honduras	1 6	Dough	8.91	3.12
Miller	10½	Ripe	10.18	2.44
Stump	1 6	Doughy	15.77	2.63
Gooseneck	1 1	Dough	13.02	2.42
White Mammoth	2	Milk	8.18	3.26
Gray Top	1 13	Milk	9.54	2.77
Neazana	1 3½	Hard dough	9.12	3.22
Early Amber	1 2	Ripe	13.08	1.96
Lynk's Hybrid	1 12	Milk	11.88	1.77
Honey	2 2	Milk	9.22	3.45
Liberian	1 6½	Ripe	11.30	2.38
Kansas Orange	1 3½	Hard dough	10.99	2.32
Early Orange	1 14	Dough	7.16	3.57
White Liberian	1 14½	Doughy	11.14	1.78
Canada Amber	12	Ripe	10.74	2.97
Texas Amber	1 2½	Ripe	13.65	2.29
Illinois Amber	1 2	Ripe	11.62	1.65

The only variety used for sugar making was the Early Amber. Three separate plots were planted, of 3.6, 2, and 1½ acres respectively. The latter plot was used for experiments with fertilizers. Each kind of fertilizer was put on a plot of cane of one-twentieth of an acre, and these plots were separated from each other by guard rows where no fertilizer was used. Each lot was cut and brought to the mill separately and a sample from the defecated juice was taken for analysis.

The following table gives the results:

Fertilizers.*	Nitrogen on plot.	Cane sugar.	Glucose.	Stripped stalks.
	Pounds.	Per cent.	Per cent.	Pounds.
No. 1. No manure..	....	10.91	2.80	828
No. 2. 2339 pounds barnyard manure	....	10.19	2.77	864
No. 3. Nitrogen mixture	1.2	10.57	2.77	796
No. 4. Superphosphate (15 pounds)	....	10.54	2.80	762
No. 5. Chloride of potassium (7½ pounds)..........	....	10.95	2.78	910
No. 6. { Nitrate of sodium (7½ pounds) } { Superphosphate (15 pounds) }	1.2	10.77	2.56	776
No. 7. { Nitrate of sodium (7½ pounds) } { Chloride of potassium (15 pounds) }	1.2	10.17	2.78	856
No. 8. { Superphosphate (15 pounds) } { Chloride of potassium (7½ pounds) }	....	10.83	2.89	778
No. 9. Barnyard manure...............................	....	10.87	2.85	578
No. 10. No manure.......................................	....	11.63	2.85	616
No. 11. Mixed minerals and nitrogen..................	1.2	11.34	2.76	472
No. 12. Mixed minerals and two-thirds nitrogen	.8	10.92	2.80	578
No. 13. Mixed minerals and one-third nitrogen	.4	10.41	2.94	548
No. 14. Mixed minerals and one-sixth nitrogen	.2	10.10	2.86	560
No. 15. Mixed minerals and one-twelfth nitrogen..........	.1	11.59	2.90	618
No. 16. Mixed minerals and no nitrogen	....	10.47	2.87	590

*These were obtained at the Connecticut Experiment Station, and a list may be found on page 363 of the Report of the Connecticut Board of Agriculture for 1880.

The method employed for making sugar was as follows: The juice from the mill was pumped into the defecator by a jet pump. To the lukewarm juice milk of lime was added to a slight alkaline reaction. It was then heated to the boiling point, and after a few minutes skimmed, again heated and skimmed in the same manner three or four successive times. The result was a very clear juice with very little sediment in the bottom of the defecator. The juice was next evaporated about 20° Beaumé in an open pan. It was then drawn into the vacuum-pan and evaporated to a rather dense sirup. From there it was drawn into wooden boxes lined with tin, each box holding fifty gallons. In a couple of days it was usually ready to separate. The table below gives the result of the work. Thus far only the "firsts" have been separated. The molasses has been concentrated and is now crystallizing quite heavily. It will undoubtedly yield at least 25-30 per cent. of sugar:

	Plot No. 1.	Plot No. 2.	Plot No. 3.
Area of plots in acres	3¾	2	1¼
Total weight of stripped cane	75,262	28,974	17,112
Per cent. of juice expressed..........................	65	47	47
Per cent. of cane sugar in juice......................	9.89	12.10	11.20
Per cent. of glucose in juice..........................	3.95	2.86	2.78
Weight of cane sugar (firsts).........................	2,116½	1,008	594
Gallons of molasses...................................	409	101	58
Yield per ton of cane Sugar (firsts).................	56.3	70	69
Yield per ton of Molasses............................	10.0	7	6.8

Plot No. 1 was on quite low and rather heavy claying soil. At least one-fourth of the heads were still in the milk. Plots 2 and 3, on the other hand, were each on a sandy slope and the cane was quite mature. This accounts for the great differences in the composition of the juices. The cane from the first three acres of plot No. 1 was crushed in a five-roller mill which submitted to four gradually increasing pressures, the rollers being so arranged that the pressure between the two last was double that between the first two rollers. The yield of juice with this mill was 69½ per cent.

I cannot give a fair estimate of the total cost of the sugar made here, as several stoppages had to be made in order to exchange some of the machinery which was partly borrowed and was called for in the midst of the work. The expenses and returns per day of twelve hours were as follows:

EXPENSES.

Fireman _____ $2 00
One man at mill _____ 1 50
Two men in cane house at $2 _____ 4 00
Half ton of coal _____ 3 25

Lime, oil, kerosine, litmus paper	$0 30
Four tons cane at $2.50 per ton	10 00
Cost of separating	2 00
Total daily expenses	23 05

RETURNS.

Two hundred and forty-five pounds sugar at 8 cents	$19 60
Thirty-six gallons sirup at 35 cents	12 50
	32 10
	23 05
Balance	9 05

The scale on which this work was conducted was, of course, entirely too small to be economical. A great deal of waste was incurred that might have been saved with better appliances. Judging by the results obtained here the past two seasons, there can be no reasonable doubt but that our sorghum cane will be a valuable crop as a sugar producing plant in this vicinity.

Very respectfully,

M. SWENSON,
Chemist Agricultural Department University of Wisconsin.

Prof. B. SILLIMAN.

34.—BIBLIOGRAPHY OF SORGHUM.

[CHRONOLOGICALLY ARRANGED.]

Arduino, Pietro. Memorie di osservazioni e di sperienze sopra la coltura e gli usi di varie piante, che servons o che servir possono utilimente alla tinctura, all'economia, all'agricoltura, etc. Tomo I, Padova, 1766, 4, xxiv, 105 p., 19 tab.

Child, David Lee. The Culture of the Beet and Manufacture of Beet Sugar. Boston, 1840. 12°, pp. 156.

Notiz über Maiszucher. Annales maratimes et coloniales. Paris, 1842. II, T. 2, pp. 346.

Colman's Rural World. Saint Louis, Mo., 1848 to 1882.

Browne, D. Jay. Researches on Sorgho Sucré by Department of Agriculture. Report 1854, p. xii and pp. 219–223.

Chinese Sugar Cane. Correspondence Department of Agriculture. Report 1855, pp. 279–285.

Barral. Ueber den Zucher in Holcus Sorghum. Moniteur industriel, 1855, p. 1939.

Reihlen. Ueber Holcus saccharatus. Polytechnisches Centralblatt, 1855, p. 703.

Wray, Leonard. Practical Sugar Planter.

Vilmorin, Louis. Paris, 1855. Essay on Sorgo Sucré in Le Bon Jardinier Almanac, pp. 41–52.

Sorghum, Characteristics of. Massachusetts Agricultural Report, 1856, (Pt. 1), pp. 89–91–98.

Sorghum Saccharatum, brought from China. Illinois Agricultural Report, 1856, 1857, p. 446.

Browne, D. Jay. Crystallization of the juice of the Sorgho Sucré. Department of Agriculture. Report, 1856, pp. 309–313.

Jackson, C. T., M. D. Chemical Researches on the Sorgho Sucré. Department of Agriculture. Report, 1856, pp. 307–309.

Erfahrungen ueber die Kultur und Ausbeute von Zucher aus Sorghum. Moniteur industriel, 1856, No. 2049.

Turrell. Ueber das Sorghum in Nord China. Moniteur industriel, 1856, No. 2110.

Zoulie. Ueber Sorghum. Moniteur industriel, 1856, No. 2110.

Madinier, P. and G. Lacoshe. Guide du cultivateur du Sorgho à Sucré. Paris, 1856.

Madinier, M. Department of Agriculture. Report, 1853, p. 313.

Vilmorin, Louis. Department of Agriculture. Report, 1856, p. 312.

Browne, D. Jay. Report of the United States Agricultural Society. Department of Agriculture. Report, 1857, pp. 181–183.

Sorghum, Experiments in the cultivation of. Massachusetts Agricultural Report, 1857 (Pt. 1), pp. 117–145, 149–215 (Pt. 2), pp. 157–222, 170, 38, 225–229–234.

Sorghum Mills, Description of. Ohio Agricultural Report, 1857, pp. 416.

Sorghum, Lovering's experiments of. Ohio Agricultural Report, 1857, pp. 423.

Sorghum, Statement of, in Ohio. Ohio Agricultural Report, 1857, pp. 437.
Hardy. Ueber Zucher aus Sorghum. Moniteur industriel, 1857, No. 2131.
Cavé. Ueber den Anbau von Sorghum auf dem Gute de Condé. Moniteur industriel, 1857, No. 2153.
Sorghum, Report on, at Fair. Ohio Agricultural Report, 1857, pp. 142.
Sorghum Sugar Cane. New York Agricultural Report, 1857, pp. 16–128–135.
Hyde, J. F. C. The Chinese Sugar Cane. New York, 1857. 12° pp.
Sorghum Saccharatum. Pennsylvania Agricultural Report, 1857–58, pp. 147–557.
Sorghum, History of. Ohio Agricultural Report, 1857, p. 409.
Sorghum, discussed in Annual Convention. Ohio Agricultural Report, 1857, p. 195.
Sorghum or Chinese Sugar Cane. Ohio Agricultural Report, 1857, p. 34.
Jackson, C. T., M. D. Chemical Researches on the Chinese and African Sugar canes. Department of Agriculture. Report, 1857, pp. 185–192.
Lovering, Joseph S. Sorghum Saccharatum or Chinese Sugar Cane. Detailed account of experiments and observations upon, 1857.
On the Identity and Hybridity of the Chinese and African Sugar Canes. (Condensed from the Proceedings of the Boston Society of Natural History.) Department of Agriculture. Report, 1857, pp. 183–185.
Smith, J. Lawrence. Investigation of the Sugar-bearing Capacity of the Chinese Sugar Cane. Department of Agriculture. Report, 1857, pp. 192–196.
Sorghum Sugar, condensed correspondence on. Department of Agriculture. Report, 1857, pp. 196–226.
Stansbury, Charles F. Chinese Sugar Cane and Sugar Making. New York, 1857. 12°.
Sorghum Syrup, statement in regard to making. New York Agricultural Report, 1858, p. 722.
Sorghum Cane, letters read at a convention on. Illinois Agricultural Report, 1858, p. 306.
Sorghum Cane. Wisconsin Agricultural Report, 1858, 1859, pp. 261–350–409–412.
Sorghum. New York Agricultural Report, 1858, p. 12.
Sorghum. Iowa Agricultural Report, 1858, p. 9.
Sorghum Cane, Sugar from. Illinois Agricultural Report, 1858, pp. 107–109.
Sorghum Cane, On. Illinois Agricultural Report, 1858, p. 512.
Olcott, Henry S. Sorgho and Imphee. The Chinese and African Sugar Canes. A treatise upon their origin, varieties, and culture. New York, 1858. 12°, pp. 352.
Jackson, Ch. F. Compte Rendus, XLVI, p. 55, 1858.
Du Feyrat, Comparative Ausbeute aus Sorghum u. Zuckerrohr. Moniteur industriel, 1858, No. 2228.
Leplay. Ueber Sorghum u. dessen Zuckergehalt. Moniteur industriel, 1858, No. 2334. Comptus rendus v. 46, p. 444. Polytechnisches Centreblatt, 1858, 593. Polytechnisches Jour. Ding. B., 148, p. 224.
Collectaneen ueber Zucher aus Sorghum. Polytechnisches Jour. Ding. B., 148, p. 158. Bulletin de la Societé d'encouragement, pour l'industriel nationale, 1858, p. 505. Polytechnisches Centreblatt, 1858, p. 1102.
Habich, Ueber Sorghum u. dessen Werth. Polytechnisches Jour. Ding. B., 148, p. 302. Polytechnisches Centralblatt, 1858. p. 1647.
Lovering, Ueber den Werth des Sorghums als Zuckerpflanze. Moniteur industriel, 1858, No. 2312. Bulletin de la Societé d'encouragement pour l'industrie nationale, 1858, p. 673.
Sorghum Saccharatum. Essay on its composition. Michigan Agricultural Report, 1859, p. 213.
Wagner. Ueber den Mais als Zuckerpflanze. Agromisches Zeitung, 1860, p. 12.
Das Zucker Sorgho, oder das chemisches Zuckerrohr (Holcus succharatus). Deutsche Gewerbezeitung Wiecks, 1859, p. 443. 1860, p. 156.
Pierre Ueber das chemische Zucker-sorgho als Futter u. Zuckerpflanze. Bulletin de la Societé d'encouragement pour l'industrie nationale, 1860, p. 94.
Anbau, versuche mit der Zuckerhirse Sorghum saccharatum. Annalen der Landwirthschaft. Wachenblatt, Berlin, 1860, p. 350.
Cook, D. M. Culture and manufacture of sugar from Sorghum. Department of Agriculture. Report, 1861, pp. 311–314.
Sorghum. Ohio Agricultural Report, 1861, p. 52.
Sorghum as an Exhauster of Soil. Ohio Agricultural Report, 1861, p. 526.
Sorghum as a Wine plant. Ohio Agricultural Report, 1861, p. 526.
Sorghum, Growth and manufacture. Ohio Agricultural Report, 1861, p. 210.
Sorghum Sugar. Warder's Statement. Ohio Agricultural Report, 1861, p. 15.
Sorghum Culture and manufacture. Ohio Agricultural Report, 1861, p. 208.
Sorghum Sugar Cane. New York Agricultural Report, 1861, p. 785.
Department of Agriculture. Report, 1861. Cornstalk Sugar, p. 275.

Sorghum Culture. Illinois Agricultural Report, 1861, 1864. pp. 553, 567.
Sorghum. Iowa Agricultural Report, 1861, p. 8.
Sorghum, History of. Ohio Agricultural Report. 1861, p. 206.
Sorghum as a forage plant. Ohio Agricultural Report. 1861, p. 527.
Sorghum Culture. Wisconsin Agricultural Report, 1861, 1865, p. 35.
Goessmann, C. A. Chinese Sugar Cane. Contributions to the knowledge of its nature,
     &c. Transactions of New York State Agricultural Society, 1861, pp. 735-811, and
     in pamphlet.
Hedges, Isaac A. Sorghum Culture and Sugar-Making. Department of Agriculture.
     Report, 1861, pp. 293-311,
Sorghum. Illinois Agricultural Report, 1861-1864. pp. 32, 67, 203, 859.
Bollman, Lewis. Cultivation of the Sorghum. Department of Agriculture. Report.
     1862, pp. 140-147.
Smith, J. H. Imphee and Sorghum Culture and Sugar and Sirup Making. Depart-
     ment of Agriculture. Report, 1862, pp. 129-140.
Statistical Report. Department of Agriculture Report, 1862, pp. 552, 553.
Sirup and Sugar manufactured from Sorghum. Ohio Agricultural Report, 1862, p. 87.
Sorghum. Iowa Agricultural Report, 1862, p. 123.
Hedges, Isaac A. Sorgho or the Northern Sugar Plant, with an introduction by William
     Clough. Cincinnati, Ohio. 1862. 12°, p. 204.
Sorghum. Ohio Agricultural Report, 1862. p. 87.
Wetherill, Charles M., M. D. Chemist Department of Agriculture. Department of Agri-
     culture. Report, 1862, pp. 514-540.
Clough, William. Sorgho Journal, Cincinnati, Ohio, 1863 to 1869.
Gould, John Stanton. Report on Sorghum. New York Agricultural Report, 1863,
     pp. 735-769.
Wallace, G. B. Sorghum. Iowa Agricultural Report, 1863, p. 162.
Moss, James W. Sorghum. Iowa Agricultural Report, 1863, p. 244.
Sorghum, Essay on. Iowa Agricultural Report, 1863. pp. 137-244.
Gould J., Stanton. Report on Sorghum and Sugar Beet Culture. Transactions of New
     York State Agricultural Society, 1863, pp. 735-769.
Sugar Evaporator and Mills. Ohio Agricultural Report, 1863, p. 100.
Sorghum Sugar, Protest from exhibitors of. Ohio Agricultural Report, 1863, p. 101.
Sorghum. Iowa Agricultural Report, 1863, p. 4.
Sugar Mills. Ohio Agricultural Report, 1863, p. 85.
Clough, William Sorghum or Northern Sugar Cane. Department of Agriculture. Report,
     1864, pp. 54-87.
Sorghum Mills. Ohio Agricultural Report, 1864.
Sorghum. Iowa Agricultural Report, 1864, p. 7.
Sorghum Mills, Report of Committee on. Ohio Agricultural Report, 1864, pp. 119, 120.
Collins, Varnum B. Sorgo or Northern Chinese Sugar Cane. Journal North China
     Branch Royal Asiatic Society, December, 1865, pp. 85-98. Shanghai, 1865.
Clough, William Production of Sugar from Sorghum or Northern Sugar Cane. Depart-
     ment of Agriculture Report. 1865, pp. 307-324.
Sorghum. Introduction into the State. Michigan Agricultural Report, 1865, p. 17.
Sorghum. Ohio Agricultural Report, 1865, p. 352 and p. 14.
Sorghum. Premiums awarded. Illinois Agricultural Report, 1865, 1866, pp. 13-97.
Sorghum, Report of Committee on. Iowa Agricultural Report, 1865, p. 225.
Tenney, A. P. Sugar Question. Iowa Agricultural Report, 1865, p. 329.
Ives, Mrs. E. F. Essay on Sorghum making. Iowa Agricultural Report, 1865, p. 225.
Peck, F. Botanical History of Sorghum. Department of Agriculture. Report, 1865,
     pp. 229-307.
Webster & Co. Sorgho Sugar Growers. The Culture and Manufacture of Sugar and
     Sirup from the Chinese and African Canes. Chicago, 1865 (?) 32°, pp. 41.
Reed, W. History of Sugar and Sugar-yielding Plants and Epitome of Processes of
     Manufacture. London, 1866.
Moser, J. Vers. St. at vol. 8, p. 93, 1866.
Jacobs Brothers. Sorgho Manufactures. Manual, &c. Columbus, Ohio, 1866.
Sorghum and its Products. Michigan Agricultural Report, 1866, pp. 169-172.
Sorghum in Delaware County. Ohio Agricultural Report, 1866, (Pt. 1), p. 193.
Sorghum and Imphee. Missouri Agricultural Report, 1866, p. 23.
Sorghum Mills. Illinois Agricultural Report, 1865, 1866, p. 202.
Sorghum, Report of Committee on. Iowa Agricultural Report, 1866, pp. 132 and 223.
Sorghum, Report of Standing Committee on. Iowa Agricultural Report, 1866, p. 223.
Sorghum Sugar Making. Ohio Agricultural Report, 1866 (pt. 2), p. 287.
Stewart, F. L. Sorghum and its products. Philadelphia, 1867. 12°, p. 240.
Sorghum and its products. Michigan Agricultural Report, 1867, pp. 65-67.

Sorghum. Report of Committee on. Iowa Agricultural Report, 1867, p. 18.
Sorghum. Resolutions and Report in regard to Michigan Agricultural Report, 1867, pp. 305–307.
Sorghum and its Products. Missouri Agricultural Report, 1867, p. 92.
Sorghum and Machinery. Awards on Ohio Agricultural Report, 1867, p. 138.
Sorghum. Ohio Agricultural Report, 1867, pp. 62, 246 (pt. 2), p. 16.
Sorghum of Van Wert County. Ohio Agricultural Report, 1867 (pt. 1), p. 158.
Sorghum and Machinery. Ohio Agricultural Report, 1868, p. 97.
Sorgho. The Journal and Farm Mechanic. Cincinnati, Ohio, February, 1869. 8°.
Sorghum. Committee on Iowa Agricultural Report, 1869, pp. 190, 195.
Sorghum, Products of. Iowa Agricultural Report, 1869, p. 17.
Sorghum, Report of Committee on. Iowa Agricultural Report, 1869, p. 179.
Sorghum, Secretary's Report on. Iowa Agricultural Report, 1869, p. 16.
Sorghum and Imphee. Wisconsin Agricultural Report, 1869, p. 27.
Sorghum. Iowa Agricultural Report, 1869, p. 16; 1870, p. 189.
Bretschneider. Notes on History of Plants and Chinese Botany. Peking, 1870.
Sorghum Sirup. Wisconsin Agricultural Report, 1870, p. 34.
Sorghum and its Products. Michigan Agricultural Report, 1870, p. 149.
Sorghum. Statement as to Iowa Agricultural Report, 1871, pp. 205 and 212.
Sorghum. Report of Secretary on Iowa Agricultural Report, 1871, p. 23.
Sorghum Sugar and Sirup. Report on Iowa Agricultural Report, 1871, p. 204.
Sorghum. Abstract of County Agricultural Societies. Iowa Agricultural Report, 1871; p. 300.
Skinner, E. W. Sorghum, Iowa Agricultural Report, 1872, p. 290.
Sorghum Sirup and Sugar, Report of Committee on. Iowa Agricultural Report, 1872, p. 325.
Caldwell, Phineas. Report of Committee on Sorghum and its Products, in Iowa Agricultural Report, 1872, p. 286.
Sorghum Plant. Nebraska Agricultural Report, 1873, p. 89.
Sorghum. Abstract of report on. Iowa Agricultural Report, 1873, p. 313.
Sorghum. The production of. Nebraska Agricultural Report, 1873, p. 89.
Sorghum. Table of Products and Acreage. Kansas Agricultural Report, 1873, pp. 89, 126.
Sorghum. Abstract of Report on. Iowa Agricultural Report, 1874, p. 304.
Sorghum. Ohio Agricultural Report, 1874, pp. 254, 636.
Basset, N. Guide Practique du Fabricant de Sucré. Paris, 1875, 3 vols. 8°.
Sorghum. Number of acres in. Georgia Agricultural Report, 1873–1875, p. 9.
Sorghum. Abstract of Report on. Iowa Agricultural Report, 1875, p. 281.
Sorghum. Tables of Product and Value. Kansas Agricultural Report, 1875, pp. 464, 469.
Sorghum. Diagram showing product and value. Kansas Agricultural Report, 1875, p. 460.
Sorghum. Georgia Agricultural Report, 1876, p. 222.
Sorghum. Abstract of Report on. Iowa Agricultural Report, 1876, p. 312.
Sorghum. Report of Committee on. Iowa Agricultural Report, 1876, p. 224.
Stewart, F. L. Maize and Sorghum as Sugar Plants. Department of Agriculture Report, 1877.
Sorghum Molasses, Gallons of. Virginia Agricultural Report, 1877, p. 43.
Sorghum. Abstract of Report on. Iowa Agricultural Report, 1877, p. 271.
Collier, Peter. Sorghum. Department of Agriculture Report, 1878, p. 98.
Maumené, E. J. Traite théoretique et practique de la Fabrication du Sucré, Paris, 1878. 2 vols. 8°.
Stewart, F. L. Sorghum Sugar made from Maize, &c. Washington, 1878.
Sorghum. Virginia Agricultural Report, 1878, p. 31.
Sorghum. Abstract of Report on. Iowa Agricultural Report, 1878, p. 386.
Sorghum. Introduction into the Country. Kentucky Agricultural Report, 1878, p. 144.
Collier, Peter. Sorghum. Department of Agriculture Report, 1879, pp. 36.
Sorghum. Abstract of Report on. Iowa Agricultural Report, 1879, p. 315.
Sorghum. Productions of. Kentucky Agricultural Report, 1879, p. 64.
Gossmann, C. A. Early Amber Cane. Report Massachusetts Agricultural College, 1879.
Sorghum. Sugar from. Vermont Agricultural Report, 1879–1880, p. 260.
Collier, Peter. Sugar from Sorghum. Vermont Agricultural Report, 1879–1880, p. 219.
Stewart, F. L. Sugar from Maize and Sorghum. Washington, D. C., 1879. 12, pp. 102.
Drummond, Victor A. W. Report on the Production of Sugar from Sorghum, 1879.

Sorghum Cane.   Proceedings of the Wisconsin Sugar Cane Growing and Manufacturing
    Association.   Wisconsin Agricultural Report, 1879–80, p. 463.
Johnston, Hon. J. W., chairman of Committee on Agriculture in United States Senate,
    on Sorghum Sugar, 1880.
Blymyer Manufacturing Co.   Sorgho Hand Book.   Cincinnati, Ohio, 1880.
Clough Refining Co.   Clough Refining Process for Sorghum, &c.   Cincinnati, 1880.
Collier, Peter.   Cornstalk and Sorghum Sugar.   Abstract of an address delivered in the
    House of Representatives.   Hartford, Conn., February 17, 1880.   Pamphlet; 8°, pp.
    23.
Collier, Peter.   Sorghum and Corn as Sugar-producing Plants.   Address delivered before
    Connecticut State Board of Agriculture at Willimantic.   1880.   Pamphlet.   8°,
    pp. 28.
Sorghum.   Sweets of Wisconsin.   Wisconsin Agricultural Report, 1880–1881, p. 331.
Sorghum.   Statistics of.   Quarterly Report Kansas Agricultural Report, 1880, pp. 21, 22.
Sorghum.   Production of.   Kentucky Agricultural Report, 1880, p. 140.
Collier, Peter.   Sorghum.   Department of Agriculture.   Report 1880, p. 37, and Special
    Report No. 33.
Ingram, W.   Sorghum Cultivation in Belooir.   London, 1880.
Department of Agriculture.   Preliminary Report.   1880, pp.
Collier, Peter.   Department of Agriculture Report, 1881, pp. 17.
Hedges, Isaac A.   Sugar Canes and their Products.   Saint Louis, Mo., 1881.
Rutger's Scientific School.   Seventeenth Annual Report, 1881, p. 63.
Tucker, J. H., Ph. D.   A Manual of Sugar Analysis, including the application in gen-
    eral of Analytical Methods to the Sugar Industry.   New York, 1881, 8°, pp. 353.
Weber and Scovell, Professors.   Sorghum.   Report on the manufacture of Sugar Sirup
    and Glucose from.   Illinois Industrial University, 1881.
Vilmovin, Andrieux.   Le Sorgho Sucré de chine et le Sorgho hatif du Minnesota, ou
    Sorgho Sucré Ambré.   Journal d'Agriculture Practique, May 8, 1880, and February
    17, 1881.
Spon's Dictionary.   Article Sorghum, London, 1881.
Ware, L. S.   A study of the various sources of sugar.   Philadelphia, 1881.
Sorghum, Sugar from.   Kentucky Agricultural Report, 1881, p. 83.
Sorghum, Cultivation and Manufacture of.   Kentucky Agricultural Report, 1881, p. 72.
Kolischet, Theo.   Sorghum, Sugar from.   Kentucky Agricultural Report, 1881, p. 86.
Locke, Wigner, and Harland.   Sugar Growing and Refining.   London, 1882.
Experimental Farm.   Madison, Wisconsin.   Experiments in Amber Cane, 1882.
Biot und Soubeiran.   Zucker in Mais, Polytechnisches Journal.   Dingler, 865213.
Pallas.   Mais Zucker Polytechnisches Journal, Ding., 945326.   Brevetes d'invention.
    Paris, T. 46, p. 146.
Vorschlag zu einer Production von Zucker aus Holcus Sorghum.   Technological Repos-
    itory, Gill, v. 10, p. 119.   Franklin Journal, 1 S., v. 1, p. 201.
Pallas.   Zucker aus Mais Polytechnisches, Journal, Ding., 635156.   Journal des con.
    usul et pract.   Paris, T. 26, pp. 97–109.
Neumann.   Zucker aus Mais Polytechnisches Journal, Ding., 67, S. 300.

○